T0143282

COMPUTATIONAL THINKING
FOR THE MODERN
PROBLEM SOLVER

CHAPMAN & HALL/CRC
TEXTBOOKS IN COMPUTING

Series Editors

John Impagliazzo
Professor Emeritus, Hofstra University

Andrew McGettrick
Department of Computer
and Information Sciences
University of Strathclyde

Aims and Scope

This series covers traditional areas of computing, as well as related technical areas, such as software engineering, artificial intelligence, computer engineering, information systems, and information technology. The series will accommodate textbooks for undergraduate and graduate students, generally adhering to worldwide curriculum standards from professional societies. The editors wish to encourage new and imaginative ideas and proposals, and are keen to help and encourage new authors. The editors welcome proposals that: provide groundbreaking and imaginative perspectives on aspects of computing; present topics in a new and exciting context; open up opportunities for emerging areas, such as multi-media, security, and mobile systems; capture new developments and applications in emerging fields of computing; and address topics that provide support for computing, such as mathematics, statistics, life and physical sciences, and business.

Published Titles

Paul Anderson, Web 2.0 and Beyond: Principles and Technologies

Henrik Bærbak Christensen, Flexible, Reliable Software: Using Patterns and Agile Development

John S. Conery, Explorations in Computing: An Introduction to Computer Science

Ted Herman, A Functional Start to Computing with Python

Pascal Hitzler, Markus Krötzsch, and Sebastian Rudolph, Foundations of Semantic Web Technologies

Mark J. Johnson, A Concise Introduction to Data Structures using Java

Mark J. Johnson, A Concise Introduction to Programming in Python

Lisa C. Kaczmarczyk, Computers and Society: Computing for Good

Mark C. Lewis, Introduction to the Art of Programming Using Scala

Bill Manaris and Andrew R. Brown, Making Music with Computers: Creative Programming in Python

Uvais Qidwai and C.H. Chen, Digital Image Processing: An Algorithmic Approach with MATLAB®

David D. Riley and Kenny A. Hunt, Computational Thinking for the Modern Problem Solver

Henry M. Walker, The Tao of Computing, Second Edition

CHAPMAN & HALL/CRC
TEXTBOOKS IN COMPUTING

COMPUTATIONAL THINKING FOR THE MODERN PROBLEM SOLVER

DAVID D. RILEY AND KENNY A. HUNT

University of Wisconsin
La Crosse, USA

 CRC Press
Taylor & Francis Group
Boca Raton   London   New York

CRC Press is an imprint of the
Taylor & Francis Group, an **informa** business

A CHAPMAN & HALL BOOK

CRC Press
Taylor & Francis Group
6000 Broken Sound Parkway NW, Suite 300
Boca Raton, FL 33487-2742

© 2014 by Taylor & Francis Group, LLC
CRC Press is an imprint of Taylor & Francis Group, an Informa business

No claim to original U.S. Government works

Printed on acid-free paper
Version Date: 20140710

International Standard Book Number-13: 978-1-4665-8777-9 (Hardback)

Visit the Taylor & Francis Web site at
http://www.taylorandfrancis.com

and the CRC Press Web site at
http://www.crcpress.com

Contents

Preface

Computational thinking is a fundamental skill for everybody, not just for computer scientists. To reading, writing, and arithmetic, we should add computational thinking to every child's analytic ability.

—JEANNETTE WING[*]

Traditionally, general education courses in computer science have been rooted in some combination of four topics: (1) computer programming, (2) computer hardware, (3) societal issues of computing, and (4) computer application skills. Computational thinking is different because the focus goes beyond introductory knowledge of computing to treat computer science as an independent body of thought that is an essential part of what it means to be educated today. Thinking algorithmically is uniquely important just as is scientific investigation, artistic creativity, or proof theory in mathematics; and yet computational thinking is a distinct form of thought, separate from these other academic disciplines. The diagrammatic techniques used in software engineering analysis are effective for such efforts as strategic planning. The way that data is digitized has a profound impact on today's graphical art and music. Computer science modeling techniques are essential in many aspects of today's research in the social sciences and business. Pattern-matching techniques are useful in even the most rudimentary forms of DNA analysis. Understanding things such as how to express software requirements and the limits of computing are essential for all people who expect to live and work in a world where information is stored, accessed, and manipulated via computer software.

This book adheres to the concept of computational thinking. Since content such as this is typically taught in more advanced computer

---

[*] Dr. Jeannette Wing is assistant director, Computer and Information Science and Engineering Directorate, National Science Foundation and former dean of the School of Computer Science at Carnegie Mellon University.

science courses, special attention is paid to the use of effective examples and analogies. In addition, every effort is made to demonstrate the ways that these concepts are applicable in other fields of endeavor and to keep this material both accessible and relevant to noncomputer science majors.

The primary topical threads of this presentation can be grouped into foundational computer science concepts and engineering topics. The foundational computer science threads include abstraction, algorithms, logic, graph theory, social issues of software, and numeric modeling. The engineering threads include execution control, problem-solving strategies, testing, and data encoding and organizing. Rather than organize all chapters around these threads, a more logically connected approach is employed. So, for example, algorithmic thinking is integral to at least six different chapters as a part of problem solving, control structures, modeling, correctness, limits of computation, and concurrency.

It is expected that anyone teaching a computational thinking course will include some instruction in computer programming. However, there are many suitable programming languages and various depths of coverage that might be appropriate. Therefore, this book does not include computer programming instruction per se. However, the fundamental concepts of programming—variables and assignment, sequential execution, selection, repetition, control abstraction, data organization, and even concurrency—are presented. Particular care has been given to present algorithms using language-independent notation.

This approach has been taught, using early manuscript versions of this book, for several semesters to university students. Reactions have been largely positive from both the students and the several faculty involved.

Authors

David Riley has been committed to computer science education for more than 35 years. He has authored eight other computer science textbooks, along with numerous book chapters and research papers. His interest in computational thinking spans countless experiences teaching computer science majors and graduate students, as well as nonmajors, and even a year as a high school teacher. He has taught a full array of computer science courses. Jeannette Wing's seminal paper, titled "Computational Thinking," and Wing's subsequent discussions at the University of Wisconsin–La Crosse caused Riley to reconsider the priorities of a computing-related education, especially as they pertain to students outside the computer science mainstream. For the past three years he has taught several sections of computational thinking to students not intending to study any other computer science. This book is based upon these experiences.

Kenny Hunt has more than 25 years of experience in the fields of computer science and engineering. His technical expertise spans a broad array of the computational spectrum: from the design of research satellite electronics to the development of large-scale cloud-based web applications. He has authored numerous research articles and published a text on image processing. He has taught computer science and software engineering to both graduate and undergraduate students for more than 15 years and is greatly intrigued by the educational benefits of computational thinking.

CHAPTER 1

What Is Computational Thinking?

Computational Thinking—It represents a universally applicable attitude and skill set everyone, not just computer scientists, should be eager to learn and use.

—JEANNETTE WING

OBJECTIVES

- To provide a working definition for the concept of computational thinking
- To introduce the distinction between analog and digital representations of data
- To examine the origins of mechanical calculation using the abacus as an example to represent, store, and process data
- To examine key historical events that contributed to the invention of modern computing hardware and software
- To explain the stored program concept and the role it plays in software execution and the manipulation of data
- To introduce the basic components and characteristics of a modern computer
- To explain Moore's law and its impact

Is there any human invention that has changed the world more than the computer? Certainly this is a question worthy of discussion. We live in a time when not owning a computer puts a person at a disadvantage in countless ways. Apart from desktop computers, laptop computers, and tablet computers, many other of today's devices rely upon embedded computers. Traction control, antilock brakes, computer-assisted parking, and even car repair all involve computers on board automobiles. Digital cameras are little more than a computer with a lens attached and most cell phones are really just handheld computers.

1.1 COMPUTERS, COMPUTERS EVERYWHERE

Computers impact nearly every aspect of life. Among the first occupations to rely upon computers were accounting and engineering, utilizing the speed and accuracy of computers for complex calculations. Later, writers, scholars, and journalists began to rely upon word processing for efficient ways to create and modify documents. Clearly, graphic artists and motion picture animators depend heavily upon computers. Consider the glass of milk you drank for breakfast. This milk most likely originated with genetically engineered crops fed to cows in rations determined by a computer chip around the cow's neck, while a computer-controlled robot milked the cows, and there were myriad computers involved with transporting, processing, and retailing the milk before you brought it home to your computer-controlled refrigerator.

Today, our finances are computer managed, our wars are fought increasingly by computer-controlled devices, and we frequently communicate with our friends through computer-reliant social networks. Unfortunately, even the fastest growing form of criminal activity is categorized as "computer crime."

The point is that you really don't have any choice about the limitless impact computing has on your life. The only choice is how to respond; you can choose to educate yourself about computers and learn to use them to your advantage, or you can choose the path of the luddite. (The word "luddite" was included in the English language not so long ago, specifically to label the person who is technology ignorant.)

1.2 COMPUTER, COMPUTER SCIENCE, AND COMPUTATIONAL THINKING

We use the terms *computer* or *computer system* to refer to a collection of computer hardware and software.* Computer *hardware* includes all of the physical devices that collectively constitute the item we think of as a

---

* Technically, it is more precise to use the term *computer system* to refer to hardware plus software, and restrict the meaning of *computer* to hardware only. However, since computer hardware is of little value without software, it is now common to use the term "computer" to mean either hardware only or hardware plus software.

desktop or a laptop computer. Such items as keyboards, LCD, computer memory, disk drives, CD and DVD drives, mice and track pads, and processors are typical parts of computer hardware.

But the computer hardware of even the most sophisticated of all computers would be of no practical value were it not for computer software. The term *software* refers to any group of computer programs. Perhaps the most important difference between a computer and other machines is the computer's ability to respond to instructions, and the instructions for performing a certain task are called a *program*. It is also acceptable to use the word *code* in place of software or program.

You have encountered numerous computer programs if you have used a computer. When you went surfing about the Internet, you were using a web browser program, such as Chrome or Internet Explorer or Firefox or Safari. Among the first things people do with a newly purchased computer is to configure the antivirus software. A computer program running on your computer might allow you to play music that was downloaded by way of a computer program running on another computer located somewhere on the Internet. Computer programs do everything from managing your bank account to formatting the pages of this book. Whenever you download any *app* to your cell phone, you have just installed a program.

Whereas most human inventions are designed to perform a specific task, computers are set apart from other machines because of the variety of tasks the computer can perform. So long as someone can create the program, the computer can perform the associated task. Often, these programs are called *applications* in recognition that the program is simply a way to apply the computer hardware to a specific purpose.

Not surprisingly, people whose career is creating programs have titles such as *programmer* or *software developer*. Since every program is designed to satisfy someone's requirements, the program is in effect solving a problem. This means that programmers are really a kind of problem solver; and given the importance of computers in our lives, computer programmers are arguably the most important of all modern problem solvers.

So how does *computer science* fit into this discussion of computer hardware and software? It turns out that study of computer science includes all issues surrounding computers from hardware to software, from the foundational theories of the technology to the end-user applications. Subfields of computer science such as computer architecture explore the way in which electrical circuits are designed, whereas software engineering examines the preferred techniques for analyzing problems, and designing and implementing programs to solve them. Some subdisciplines of computer science, like graphics, robotics, information security, networking, and artificial intelligence, study the concepts implied by their names. All of these computer science topics, and others, play a role in this book.

The preceding discussion has been leading to the central issue of this book, namely, *computational thinking*. The best way to characterize computational thinking is as the way that computer scientists think, the manner in which they reason (Figure 1.1).

Of course it is not possible to explore everything that is known to computer science. So we have selected computer science concepts, techniques, and methods that have the widest utility to those individuals who most likely will not be computer scientists. In other words, this is written to capture how computer scientists think for the rest of us. Some of the book's topics are necessary simply to be literate in a society that is so dependent upon computers. Some of the concepts will allow you to more effectively use computers in your own field. Many of these ideas are borrowed from more advanced computer science courses. However, it is increasingly the case that computing concepts are used everywhere. Words like "multitasking," "downloading," and "flash memory" illustrate how computer science jargon has found its way into everyday speech. Discoveries in many fields would not have been possible without computers. Human genome sequencing requires the processing of thousands of genes made from billions of base pairs. Motion pictures rely on computational techniques, such as wire frame models, to create lifelike images of fictional

Computational   Thinking?

FIGURE 1.1   Computational thinking?

worlds. Modern medicine is practiced with minimal invasiveness due to robotics.

The scientific community discovered roughly a decade ago that most future scientific discovery would require computing knowledge among the researchers. As a result, new specialties, such as computational biology and computational physics, have become common in institutions of higher education.

But computational thinking is useful well beyond the scientific community. A computing subfield known as "artificial intelligence" has led to significant discoveries in psychology. Many software engineering tools used in software design have proven to be highly effective as business management tools. Computer programs have revolutionized how music is written, and architects use computer imagery to visually "walk about" buildings long before they are built. In short, computers allow us to study things that were previously too small, too large, too distant, too fast, or too complex. But as every good carpenter knows, you cannot get the most out of a tool unless you know how to use it.

1.3 FROM ABACUS TO MACHINE

We begin the history leading up to modern computers by considering calculating devices, because an important aspect of computer hardware is the ability to perform calculations. Certainly, the earliest known calculating device is the *abacus*. Although it is believed that the abacus was used in Mesopotamia centuries before, the oldest archaeological evidence of an abacus dates back to approximately the fifth century BC and the oldest known written description of an abacus is estimated to have been written in China in the thirteenth century AD.

There are variations on the basic structure of this device; we shall examine a version most commonly used in recent years and known as the "Chinese abacus" (see Figure 1.2).

The abacus consists of beads strung onto spindles. Each spindle is supported from its ends, as well as through a bar offset from its center. The number of spindles can differ from one abacus to another. The bar separates the beads into two groups. The key thing to remember while using an abacus is that every bead should be pushed as far as possible toward one end of the spindle or the other. In other words, no bead should ever be positioned to allow more than one bare space (region of exposed spindle) on each side of the bar.

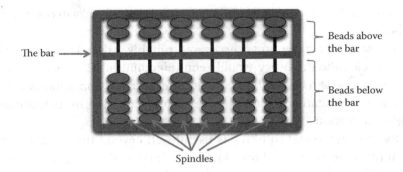

FIGURE 1.2 The Chinese abacus.

The beads have values that increase right to left, just like the value of digits in a decimal number or a Roman numeral have increasing value from right to left. The rightmost spindle of beads below the bar are called the 1s beads because each has a value of 1, while the beads of the rightmost spindle above the bar are the 5s beads. For the second spindle from the right below the bar are 10s beads and above the bar are 50s beads. The third bar from the right has 100s beads below and 500s beads above, and so forth.

Only the beads pushed as close as possible to the bar contribute to the value. This means that to make the abacus represent the value 4 you should push four 1s beads against the bar and all other beads away from the bar. Figure 1.3 illustrates both the value 4 and the value 2,639 as they could be represented on an abacus.

Different kinds of abacus may have different numbers of beads on each spindle, but the Chinese abacus always has two beads above the bar and five below. This configuration allows the abacus to represent most numbers in multiple ways. For example, Figure 1.4 shows three different ways to represent the number 10.

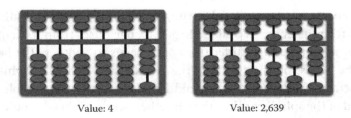

Value: 4   Value: 2,639

FIGURE 1.3 Two abacus configurations and their values.

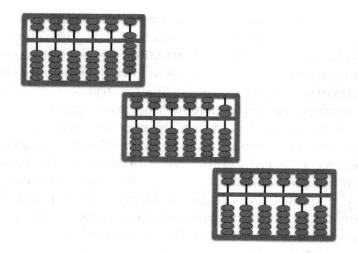

FIGURE 1.4   Three ways to represent 10 with a Chinese abacus.

Modern computers borrow four concepts from the abacus:

1. Storage

2. Representation

3. Calculation

4. User interface

For any valid bead configuration we can think of the abacus as storing the associated numeric value. So long as the beads are not moved, the abacus retains this same numeric value. A significant aspect of a modern computer is its storage. Of course, your computer can store much more than a single number, and it does not use beads, but both the abacus and your computer are definitely capable of storage.

If there is storage, then there must be something to store. The items that are stored are commonly referred to as *data*. An abacus can only store a single datum at any point in time, while your laptop can store trillions of pieces of data.

The second concept your computer borrows from the abacus is the notion of *representation*. A representation occurs anytime the data from one system is intended to model something else (the information being represented). The abacus stores (represents) an integer, using beads on a spindle to do so. The location of beads can be translated into a numeric value—the value that is represented. A modern computer is designed to solve problems that involve real-world information. That information is represented as data within computers using various technologies, many of them electronic. The electronic signals inside your computers memory can be translated into the information that they represent. We will discuss just how computers represent information in more detail in Chapter 2.

The third property of a computer also present in an abacus is the ability to perform calculations. Truthfully, neither the abacus nor computer hardware alone can perform calculations. In the case of the abacus something (usually a human) must push the beads around. Addition and subtraction are possible by adding or removing beads next to the bar. As mentioned before, computer hardware also requires something, namely, software, in order to perform calculations. Just like humans can cause an abacus to perform arithmetic, software can cause computers to perform computations.

As a final similarity to modern computers, the abacus illustrates the first known user interface for a calculating device. The term *user interface* refers to the way that humans communicate with the machine. In the case of the abacus the user interface consists of the use of fingers and thumbs to slide beads mounted on spindles and to visually interpret the represented value by the location of beads. The user interface on your laptop computer is much more sophisticated, using a keyboard and a trackpad together with some kind of liquid crystal display (LCD). We say that you use a *graphical user interface (GUI)* because most computer interaction involves the manipulation of graphical images, such as icons, buttons, sliders, pop-up windows, and pull-down menus.

The abacus may exhibit some concepts still in use by today's computers, but no one would use the word "computer" to describe an abacus. The abacus does not have enough storage, is designed to represent only integers, is limited in the kind of calculations it can perform, and has a rather crude user interface.

The importance of improving the calculation capabilities of human inventions was evident for many centuries after the abacus. One example device of note was *Napier's bones* invented by a Scottish mathematician named *John Napier* and published in 1617. Napier's bones consist of small rectangular sticks with numbers and lines on each stick. Different sticks have numbers positioned in cleverly different ways (Figure 1.5). Arranging the sticks in different ways makes it convenient to perform multiplication, division, and even calculating square roots.

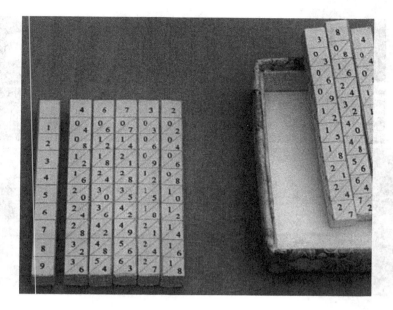

FIGURE 1.5 Napier's bones.

The next step toward computer hardware improved both the speed of calculation and user interface, while bringing human invention into the category of something that could validly be called a "machine." Actually, there were a few inventions that occurred in history during roughly the same time. These first calculating machines were invented by mathematicians and from various countries in Europe. Perhaps the two most significant of the earliest mechanical calculators were *Pascaline*, invented in 1643 by Frenchman *Blaise Pascal*, and *Leibniz' calculator* invented by the German mathematician and philosopher *Gottfried Leibniz* around 1674. Figure 1.6 and Figure 1.7 show photos of these machines.

Pascaline and the Leibniz' calculator advanced the user interface by permitting the user to turn cranks and thumb wheels. These devices also did a better job of assisting humans through the use of internal wheels, gears, and levers that accomplished addition, subtraction, multiplication, or division. These machines also demonstrate the importance of speed when performing calculations. Presumably, a knowledgeable user could perform lengthy calculations more rapidly using these machines rather than an abacus or Napier's bones. These were early devices that already illustrated man's conquest of finding machines capable of accelerating calculations.

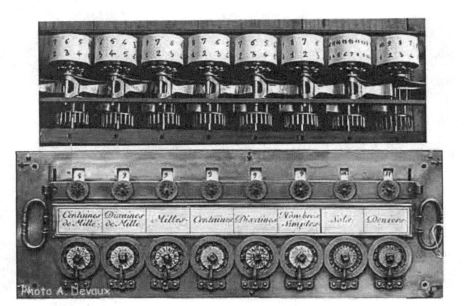

FIGURE 1.6   Pascaline.

FIGURE 1.7   Leibniz' calculator.

FIGURE 1.3   Fragment of Antikythera mechanism.

Not every historical calculating device was used to perform arithmetic. The device that is the first known example of using gears for calculation is the *Antikythera mechanism* (Figure 1.8). Dated to the first century BC, this machine contained at least 30 interconnected brass gears of various dimensions. It is believed that positioning a crank on the Antikythera mechanism caused the device to accurately identify the location of the sun, moon, and planets.

The Antikythera mechanism has been called a computer by some people. However, it is probably more accurate to think of it as a special purpose calculator, somewhat related to time-keeping machines. Remarkably, fifteen to sixteen centuries would pass before the gearing technology of the Antikythera mechanism would reappear in the watch-making industry and early calculators, such as Pascaline.

1.4 THE FIRST SOFTWARE

None of the devices described in Section 1.3 were truly *programmable*. Yes, it is possible to rearrange beads, relocate bones, or turn wheels and cranks, but these are merely ways to configure devices to perform calculations. In order to perform a different calculation any prior configuration is lost. A truly programmable device is one in which the program is divorced from the hardware so that it can be stored for reuse at a different time. In other words the program "instructs" the device in how to perform, and different programs produce different results.

The first known programmable machine is not a calculator; it is a loom for weaving cloth. Around 1805, a French inventor named *Joseph-Marie Jacquard* built the first known programmable machine. The *Jacquard loom* (Figure 1.9) was similar to other looms of the day except that it used a loop of stiff paper cards as a program. The cards had holes punched in them. Changing the number and placement of holes in these cards would cause the loom to weave a different pattern. The loom was built so that the loop and cards could be removed and replaced by a different loop of cards; thereby programming the loom to weave different patterns. This kind of punch card program is still used on textile looms today.

Although punched cards might represent programmability, weaving on a loom is quite different from computer-like calculations. The first example of what might be termed "computer software" (or at least calculator software) did not occur until approximately 1843. This important event in history came from an English mathematician and inventor named

FIGURE 1.9 Model of a Jacquard loom.

FIGURE 1.10   A piece of the Analytical Engine.

Charles Babbage. Babbage had already built a mechanical calculator capable of more advanced logarithmic and trigonometric calculations, but he did not add the notion of programmability until the design of his second, and more significant, invention—the *Analytical Engine* (Figure 1.10).

The Analytical Engine adopted the concept of *punched cards* to store and input a program into the hardware. But programs for the Analytical Engine were capable of performing a sequence of mathematical operations in the same way that modern computers can perform complex mathematical operations as directed by a proper computer program. Sadly, because of the complexity of the device, the manufacturing capabilities of the day made it impossible to construct a complete Analytical Engine during Babbage's lifetime.

An interesting side note in history often told about the Analytical Engine involves a woman named *Ada Lovelace* (Figure 1.11). The Countess Lovelace was the daughter of the famous poet, Lord Byron. She was quite interested in the work of Charles Babbage and is known to have written programs for the Analytical Engine. Some people have called Ada Lovelace the first programmer, but this cannot be confirmed and is most likely not true, since several individuals (Babbage included) wrote programs at about the same time. Nonetheless she is clearly among the first programmers.

FIGURE 1.11 Charles Babbage and Ada Lovelace were among the first computer programmers.

1.5 WHAT MAKES IT A MODERN COMPUTER?

One widely accepted definition of *modern computer* requires three properties of this calculating device:

1. It must be *electronic* and not exclusively mechanical.

2. It must be *digital* and not analog.

3. It must employ the *stored program concept*.

As it happens, even the Analytical Engine invented by Babbage fails to satisfy every one of these three requirements.

To find the first invention that is believed to satisfy at least one of these three properties, we skip to the 1890s. The United States has a long history of taking census every ten years. In 1880 the census was tabulated, like every decade prior, by hand. This process of counting citizenry and categorizing them by geographic region was becoming difficult because of rapid population growth. In fact the 1880 census was barely completed before 1890 when the next census was to begin.

A man named *Herman Hollerith* invented a calculating device built specifically for tabulating the US census. Hollerith's machine completed the 1890 census in less than one year. More important for computing, Hollerith's machine ran on electricity. The Hollerith tabulating machine can fairly be labeled as the first calculating (i.e., computer-like) hardware that satisfies any of the properties that distinguish a modern computer.

Hollerith later founded a Tabulating Machine Company to build these devices, and his company merged to form IBM Corporation in 1924; IBM remains today as one of the world's largest manufacturers of computers. Hollerith's tabulating machine also provides convincing evidence of the future capacity of computers to assist in solving human problems.

Before revealing the candidates for the first modern computer, there are two of the preceding properties of a modern computer that have yet to be mentioned. The first issue is that a modern computer must be digital. Prior to the 1930s, machines that stored data typically did so as represented using mechanical gears or electrical signals. Gears can generally be rotated to an infinite number of different angles. Similarly, electrical signals are infinitely variable in terms of voltage, amperage, capacitance, and inductance. This kind of continuous change is called *analog*. For example, an analog wristwatch often has a sweep second hand and can position the minute hand at an infinite number of positions around the dial.

A *digital* system, unlike analog systems, is one in which there are not an infinite number of possibilities and change is not continuous. Instead, digital systems restrict values to be one of a few choices. For example, hours and minutes on a digital watch are displayed as numbers. It is not possible for the minute number to display anything between 9:30 and 9:31. Most of our automobile speedometers are analog with a needle that rotates gradually as the car accelerates. However, a few cars have digital speedometers that display the current speed as a single number in either whole miles or meters per hour.

An explanation of the stored program concept requires a brief look at the major units of hardware in a modern computer. Figure 1.12 diagrams a simple desktop-style computer with three components: a keyboard, a display, and a system unit. These three components can be used to illustrate

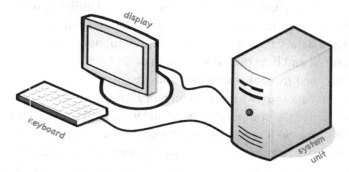

FIGURE 1.12   The basic parts of a simple desktop computer.

the four essential parts of a computer. Every modern computer must have at least one of each of the following:

1. *Input* device

2. *Output* device

3. *Memory*

4. *Processor*

An input device, as its names implies, provides a way to get data into the computer. The input device shown in Figure 1.12 is a keyboard. Other kinds of input devices include computer mice, track pads, and microphones. Smartphones and tablet computers use the surface of the LCD as an input device.

Output devices provide ways for the computer to share the results of its computation with the user. On personal computers the most common output device is the computer display. Other output devices include printers and speakers. Together, input devices and output devices provide the computer hardware that supports the user interface.

Some devices are connected to computers in order to support both input and output. For example, the data (i.e., the images) for viewing your favorite motion picture may be input to your computer from a DVD player/recorder, but that same DVD device can be used for output when you archive the photo collection on your computer. The term *I/O*, pronounced "eye oh," is commonly used by computer scientists to refer to the combination of input and output. Disk drives, flash memory cards, CD units, and DVD units are all examples of I/O devices.

Data that is input needs to be stored. Such storage occurs within the system unit in a component known as *memory*. Unlike human memory, the memory chips inside a computer almost never fail to correctly store and retrieve data. Chapter 2 explores more about how computer memories are measured and how they store data in digital form.

You've typed in your user name and password, and the computer has stored these data somewhere in its memory. One more thing needs to happen in order for the computer to be of any value—the computer must *process* your input. So a processor is the fourth essential part of a computer. Like the mechanical calculators from the seventeenth century, computer processors can perform numeric calculations; but the range of potential computer calculations clearly goes far beyond these early machines. In addition, today's computer processors typically perform trillions of operations per second.

Computer memory and computer processors are less visible than I/O devices. Generally, memory and processors consist of small integrated circuits approximately 1 cm². Both memory and processor are located within the system unit box of Figure 1.12. However, the fact that integrated circuits are so small also makes it possible to locate them more compactly. For example, the memory and processor(s) of a laptop are positioned in the same box just underneath the keyboard. Some desktop computers place the processor(s) and memory inside the same case as the display. Tablet computers and smartphones package all four components—input device, output device, memory, and processor(s)—in a single case.

You know that computer data is stored in the computer's memory and is manipulated by the computer's processor, but how is the processor instructed regarding the particular calculations to perform? In other words, how does computer software (the program and the instructions) work together with the computer hardware components? The answer is that computer processors respond to particular instructions, known as *machine instructions*. Different processors respond to different machine instructions, just like different human cultures use different natural languages.

Input devices can supply the instructions to the computer. A user who knows the instructions could input them via the keyboard, or the software could be downloaded (input) from the Internet. The program will not be available to the processor until it has been loaded (moved) into computer memory. This is known as the *stored program concept*—the third, and last, requirement of a modern computer. The point is that computer memories today are used for two things: (1) they store data, and (2) they store the instructions that process that data. The stored program concept also requires circuitry for the computer to transfer instructions from memory to the processor so that they can be executed. Typically, the processor not only executes the instructions but also controls their retrieval from memory.

1.6 THE FIRST MODERN COMPUTER

Section 1.5 defines a modern computer to be a calculating machine that is (1) electronic, (2) digital, and (3) employs the stored program concept. In this section we examine why it is not so easy to determine which invention actually qualifies as the very first modern computer.

The transition from mechanical calculators to electronic computers took place most significantly with machines known as *differential analyzers*. Differential analyzers were devices designed to solve differential equations, and mechanical versions of these calculators can be traced back to the early to mid-1800s. In 1912 a mechanical differential calculator powered by electricity was designed for use in British naval gunnery. In 1931 Vannevar Bush of the Massachusetts Institute of Technology published a journal paper describing how to construct a differential analyzer that was basically an electronic, as opposed to a largely mechanical, device [1]. However, all of these earlier devices were not modern computers, because those that used electricity did so with analog circuitry.

Influenced by these and other work on differential analyzers, two researchers from the University of Pennsylvania—*John Mauchly*, a physicist, and *Peter Eckert*, an electrical engineer—conducted a project that completed *ENIAC* (Electronic Numerical Integrator and Computer) for the US Army in 1946. In 1947 a patent was filed with the US Patent Office that declared ENIAC to be the "first electronic computer." Figure 1.13 is a photograph of ENIAC.

FIGURE 1.13   ENIAC—an early modern computer.

Clearly, ENIAC satisfies the earlier criteria for a modern computer. It was a calculating device that ran on electricity and was digital. The original version of ENIAC did not truly follow the stored program concept, but this capability was later incorporated. By today's standards ENIAC was enormous, physically filling an entire room. Its circuitry relied on 19,000 vacuum tubes (see Figure 1.14) and 1,000 relays. Vacuum tubes can be used as memory devices, but they are large (each roughly the size of a human thumb) and unreliable relative to today's computer memory. Relays are a form of electrical switch that are mechanical and also large (about the size of half a cell phone).

Despite the patent, ENIAC is not considered the first modern computer. In fact the US Patent Office invalidated the 1947 patent in 1973. The primary reason for this invalidation was the discovery of some earlier work that was not fully patented.

In 1937–1938 two physicists—*John Atanasoff* and *Chuck Berry*—at Iowa State University built a machine they called the *ABC Computer* (see Figure 1.15). During the patent dispute, it was discovered that the *Des Moines Register* had printed an article regarding the ABC Computer in 1941. It was also claimed that Atanasoff discussed his design with Mauchly in 1940 and had visited a US Patent Office that same year.

Unfortunately, the ABC Computer may not truly qualify as the first modern computer, because it failed to use the stored program concept, nor was it truly programmable for general purposes, as it was only designed to

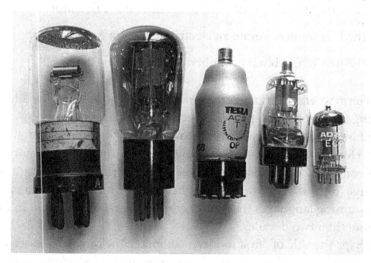

FIGURE 1.14 Vacuum tubes.

FIGURE 1.15 ABC Computer.

solve systems of linear equations. Still, the ABC Computer is considered to be first in three important ways:

1. The first fully electronic and programmable calculator

2. The first to incorporate an electronic memory

3. The first to use binary numbers (explained in Chapter 2)

A German engineer, named *Konrad Zuse,* is sometimes credited with creating the first general-purpose electronic digital computer. It was called *Z4* and built in 1945. Z4 was also the world's first commercial digital computer. However, like the ABC Computer, Z4 did not use the stored program concept. Zuse is also well known as the creator of a programming language, known as Plankakul, that became the forerunner of a family of programming languages that dominated software development in Europe for more than two decades.

Perhaps the title of "first modern computer" belongs to a lesser known invention: the *Manchester Small-Scale Experimental Machine,* also known as *SSEM* or Baby. SSEM was finished in June 1948 at Victoria University of Manchester, England, and the first computer to use the stored program concept. However, it was never intended to be a practical computer but rather part of a test bed for other hardware. By September of the same year, ENIAC had been modified to use stored programs, making it a contender for first modern computer.

Regardless of which invention should be considered to be most significant, what is clear is that during the late 1930s and throughout the 1940s there was a flurry of research taking place around the world to create early computing devices.

Within a few years computer scientists had grown weary of the tedious activity of using machine instructions, which led to the invention of *high-level programming languages*. A high-level language is one that relies upon instructions that are much more English-like instead of the cryptic numeric form of most machine instructions. The revolution leading to computer systems like today's changed to more of an evolutionary history by the mid-1950s, except for one major change to the hardware.

1.7 MOORE'S LAW

No discussion of today's computer hardware would be complete without the inclusion of one more discovery. In the 1950s and 1960s several physicists, most notably *Jack Kilby* and *Robert Noyce*, were working on a technology that would soon replace the use of vacuum tubes and relays with smaller, faster, and far more reliable electronics.

The idea was to use silicon wafers that are manufactured in such a way that thousands, and later trillions, of electronic switches, known as "transistors," can be combined onto a single chip. Such devices are referred to as *integrated circuits* and the technology that permits silicon to function in this way is called *semiconductor technology*.

Figure 1.16 is a photograph of an integrated circuit. The blackened square region in the center is the silicon wafer with wires (the lines) connecting it to the metal legs on the outside of the device. These legs typically plug into a socket for connection to the remainder of the computer circuitry. The entire package is commonly referred to as a *chip*.

Robert Noyce was awarded the Nobel Prize in Physics for his work in the creation of semiconductors. Together with *Gordon Moore*, he founded Intel Corporation—the largest manufacturer of computer processors in the world.

Integrated circuits make it possible for us to carry computers in a briefcase or a pocket that are millions of times faster than the room-sized

FIGURE 1.16   An integrated circuit.

computers of the 1950s and 1960s. Gordon Moore observed this performance trend in a paper he published in 1965 [2]. What has now become famously known as *Moore's law* was a prediction that manufacturing capabilities would advance so that the number of components within an integrated circuit would roughly double every 18 months. Moore's law has been amazingly accurate for over 40 years, explaining why computers have grown so small that they are now embedded in countless devices from wristwatches to MP3 players.

Doubling over a fixed time is a growth rate, known to mathematicians and computer scientists as *exponential growth*. In this case the growth in an exponent (or power) of 2. Following is a table showing how this exponential growth works for a base of 2:

After 18 months	Improvement factor = 2^1 = 2
After 3 years	Improvement factor = 2^2 = 2 × 2 = 4
After 4.5 years	Improvement factor = 2^3 = 2 × 2 × 2 = 8
After 6 years	Improvement factor = 2^4 = 2 × 2 × 2 × 2 = 16
After 7.5 years	Improvement factor = 2^5 = 2 × 2 × 2 × 2 × 2 = 32
After 9 years	Improvement factor = 2^6 = 2 × 2 × 2 × 2 × 2 × 2 = 64
After 10.5 years	Improvement factor = 2^7 = 2 × 2 × 2 × 2 × 2 × 2 × 2 = 128
After 12 years	Improvement factor = 2^8 = 2 × 2 × 2 × 2 × 2 × 2 × 2 × 2 = 256
After 13.5 years	Improvement factor = 2^9 = 2 × 2 × 2 × 2 × 2 × 2 × 2 × 2 × 2 = 512
After 15 years	Improvement factor = 2^{10} = 2 × 2 × 2 × 2 × 2 × 2 × 2 × 2 × 2 × 2 = 1,024

An important side effect of Moore's law has been that as more components are squeezed into the same integrated circuit space, electricity travels a shorter distance. This means that the circuits are faster. This rapid increase in the speed of our computers, often considered to be a corollary of Moore's law, is taken for granted today. In fact our computers have been doubling in speed roughly every 1.5 years for several decades. This explains why the computer purchased by a college freshman is roughly eight times less capable than those available four and a half years later. Moore's law can also be used to explain why computers have increased in speed and capacity by about 1,000 times in the past decade and a half. Chapter 10 examines a bit more about Moore's law and its implications in the future.

1.8 SUMMARY

The thinking of computer scientists is often dictated by the technologies they employ. This chapter explores the basic inventions that led to the computer hardware and software in use today. It is interesting to note that during the last 50 years arguably the most significant thing to happen with computers is their ability to solve ever more problems. This increase in computer application is a direct result of Moore's law, resulting in computers that are smaller, faster, and cheaper.

1.9 WHEN WILL YOU EVER USE THIS STUFF?

Two things explain why computers have become so prominent in our modern world: (1) we have discovered ways to store and manipulate information via computers and (2) we have created extremely fast computers. Moore's law describes the pace of change in computing history—a pace unmatched by other human inventions. No one can pinpoint the future of human discovery, but based upon history computers are likely to play a pivotal role in whatever the future brings.

REFERENCES

1. Bush, V. "The Differential Analyzer: A New Machine for Solving Differential Equations." *Journal of the Franklin Institute* 212, no. 4 (1931): 447–488.
2. Moore, Gordon E. "Cramming More Components onto Integrated Circuits." *Electronics Magazine*, April 19, 1965, vol. 38, no. 8, p. 4.

TERMINOLOGY

abacus

ABC Computer

analog

Analytical Engine

Antikythera device

app

application

Attanasoff, John

Babbage, Charles

Berry, Chuck

calculation

chip

code

computational thinking

computer

computer science

computer system

data

differential analyzer

digital

Ekert, Peter

ENIAC

electronic (computer)

exponential growth

graphical user interface (GUI)

hardware

high-level programming language

Hollerith, Herman

I/O

input

integrated circuit

Jacquard, Joseph-Marie

Jacquard loom

Kilby, Jack

Leibniz, Gottfried

Lovelace, Ada

machine instruction

modern computer

Moore, Gordon

Moore's law

Napier, John

Napier's bones

Noyce, Robert

output

Manchester Small-Scale Experimental Machine (SSEM)

Mauchly, John

memory	semiconductor technology
Pascal, Blaise	software
Pascaline	software developer
processor	stored program concept
program	storage
programmable	user interface
programmer	Z4 Computer
punched cards	Zuse, Konrad
representation	

EXERCISES

1. What are the three qualities required of a calculating device in order for it to qualify as a modern computer?

2. What is the difference between computer hardware and computer software?

3. Describe the significance of each of the following inventions, as it eventually led to the creation of the first modern computer.

   a. The abacus

   b. The Analytical Engine

   c. Jacquard loom

   d. Hollerith's machine for tabulating the US census

4. Digital cameras are one kind of modern computer. In this sense, answer the following questions.

   a. How does the user supply *input* to a digital camera?

   b. What would you consider to be the camera's *output* device(s)?

   c. What is the purpose of the camera's memory?

5. In what sense might you consider a portable music player to be *programmable*?

6. In what ways do the GUI elements of a smartphone or tablet computer typically differ from those of a laptop or desktop computer?

7. Applying Moore's law, how much larger would you expect a computer from 30 years ago to be by comparison to the computer you currently use?

CHAPTER 2

How Real-World Information Becomes Computable Data

It is a capital mistake to theorize before one has data.

—ARTHUR CONAN DOYLE

OBJECTIVES

- To define the terms *information* and *data*
- To describe how data is encoded as bit strings within a computing system
- To define measures of data capacity and how much capacity is required to store certain types of real-world information
- To describe positional numeral systems
- To describe how integer and real numbers can be encoded
- To show that computers can be imprecise
- To describe how complex information such as text, colors, pictures, and sound can be encoded as bit strings

Since computer programs process real-world information, the first step of computational problem solving is to encode real-world information as data that can be processed by a computer. Real-world information comes from many sources and in a variety of forms. Converting information into data that a computer can store and understand presents many challenges. For example, how can an audio recording, or a seventeenth-century Dutch oil painting or a full-color page from a textbook or even a fingerprint be stored inside of a computer?

This chapter will describe the underlying encoding system used by most computing systems and will show how complex types of information such as images and sound can be encoded by using this encoding system. This chapter will also discuss the techniques for minimizing the encoding of information such that the data is more compact and efficient.

2.1 INFORMATION AND DATA

The Industrial Revolution was a period of time where extensive changes in manufacturing and technology brought about equally extensive changes to the global economic and cultural environment. Starting in the late 1800s, the economies of Europe and later the United States began to move from a largely manual and animal-based agricultural system toward a machine-based manufacturing system. Commerce and trade expanded greatly due to the construction of canals, roads, and especially railways. The revolution was fueled primarily by steam, coal, and by the development of iron and metal machines for industrial use. Almost every aspect of daily life was affected as more sophisticated technologies and equipment were used throughout society.

Recent times have seen another tremendous upheaval in the global economic and cultural environment. The major economies of our time have shifted from a primarily industrial base toward an economy that is based on the generation, capture, and analysis of digital information. The resulting social and economic climate is known as the Information Age; a period of time that is characterized by massive amounts of digital data that are easily accessible, quickly transmitted, and subject to meaningful analysis. The transition to an information-based economy and culture was propelled primarily by advances in electronic computing and networking. Personal computers gained widespread adoption in the late 1970s and this, coupled with the rapid development of the Internet in the mid-1990s, served to bring digital data and digital computation into the cultural mainstream.

The terms *information* and *data* are largely overlapping ideas, but in this book we will make a distinction between them. The *Oxford English Dictionary* (OED) defines *data* as "the quantities, characters, or symbols on which operations are performed by computers and other automatic equipment, and which may be stored or transmitted in the form of electrical signals, records on magnetic tape or punched cards, etc." The OED also makes a distinction between information and data when it defines *information* as "knowledge communicated concerning some particular fact, subject, or event; that of which one is apprised or told; intelligence, news"

FIGURE 2.1   A two-dimensional encoding of the phrase "Computational think-ing is for everyone."

and adds that information is "that which is obtained by the processing of data." Throughout this book we will use the term "information" to denote a fact or piece of knowledge about the real world, whereas we will use the term "data" to denote an encoding of information such that the information can manipulated by a computing system.

Consider the following examples that distinguish between information and data. Your credit card number is information that is turned into data when it is stored on the back of your card. The encoding of this information may take the form of a strip of magnetically charged particles, a one-dimensional bar code, a two-dimensional bar code, or even as a pattern of etchings in a radio-frequency identification (RFID) chip. Your name is also a piece of information that is turned into data when it is stored in a computing system. The encoding may be electrical charges stored on a digital circuit or even as a one- or two-dimensional bar code on the back of an identification badge. The idea that "computational thinking is for everyone" is a piece of information that can also be turned into data by encoding. Figure 2.1, for example, shows a two-dimensional encoding of this very phrase.

2.2 CONVERTING INFORMATION INTO DATA

One of the challenges of the information age is the transformation of real-world information into data. How can a painting such as the Mona Lisa be stored in a computer? How can an orchestral performance by the London Philharmonic Orchestra be stored on a computer and transmitted around the globe? How can your voice be recorded as data on a voicemail system or how can your smartphone's password be stored as electronic data within the system?

We first note that there are two different types of data: *continuous* and *discrete*. Data is continuous if there are an infinite number of possible values for an individual datum, whereas data is discrete if there are a finite number of possible values. Continuous data is usually associated with measurements involving the physical or real world, whereas discrete data is usually associated with things that can be counted.

As an example, consider measuring the weight of an orange. Although the orange could weigh exactly 200 grams, it might also weigh 229.3 grams or 229.31533 grams or even 229.31533480185993 grams. In other words, the weight of an orange is an example of continuous data since there are an infinite number of possible values that might describe the weight of an orange. On the other hand, consider asking your friends how many biological parents they have who are still living. They may respond with the numbers 0 or 1 or 2. Since no person has more than 2 biological parents, numbers larger than 2 are not possible and it should be obvious that it is not possible to have a fractional number of living parents. The number of living parents is an example of discrete data since there are only a finite number of values that describe this situation.

In electronics, a signal may be either *analog* or *digital*. An *analog* signal is an encoding of continuous data, whereas a digital signal is an encoding of discrete data. In earlier times, electronics were systems that processed analog signals, but modern computing systems almost exclusively utilize digital electronics. The reason that digital systems are generally preferred to analog systems is that there are a finite set of possible values to process in a digital system.

In digital systems, the smallest unit of data is known as a *binary digit*, or *bit*. At any point in time, a bit can only take on one of two possible values: ON or OFF. You can think of a bit as an extremely small battery that can

| 0 | 0 | 0 | 1 | 1 | 0 | 1 | 1 |

FIGURE 2.2   The four patterns that a string of two bits can exhibit are 00, 01, 10, and 11.

be very quickly charged or discharged. When charged, the bit is ON and when discharged, the bit is OFF. Mathematically speaking, a bit is usually denoted as the value 0 when OFF and the value 1 when ON. Determining whether a bit is ON or OFF is straightforward. Consider, for example, how easy it is to determine wither a lightbulb is on or whether an electric fence is off! Throughout the remainder of this text, we will denote the OFF state as a 0 and the ON state as a 1.

You might be surprised to discover that computing systems encode all information as a sequence of bits. Pictures, sound, textbooks, and video are encoded as long sequences of bits. A sequence of bits is commonly referred to as a *bit string*. Since the bits in the string are able to vary in the values that they hold, some bits being 1 while other bits are 0, a bit string is able to display a great number of different patterns. For a single bit (i.e., a bit string of length one), there are only two patterns that the string could exhibit at any one point in time. The bit could either be 0 or 1. Consider, however, a string of two lightbulbs. How many different patterns could the string exhibit? Two of the patterns are obvious: both lightbulbs could be 0 or both lightbulbs could be 1. Two other patterns are also possible. The first lightbulb could be 0 and the second lightbulb could be 1. It is also possible that the first lightbulb could be 1 and the second lightbulb could be 0. These four patterns are illustrated in Figure 2.2 where lightbulbs are used to depict a single bit.

Now consider longer bit strings. How many different patterns can a bit string of length three exhibit at any one point in time? If the additional bit is placed as the first bit in the string, it is easy to see that there are twice as many patterns as before. If the first bit is 0 then there are four patterns that the remaining bits can exhibit. If the first bit is 1 (the only other possibility for that bit) there are four patterns that the remaining bits can take on. We conclude that there are eight patterns that a bit string of length three can exhibit at any one point in time. This is illustrated in Figure 2.3.

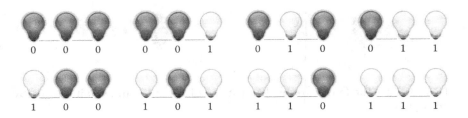

0 0 0   0 0 1   0 1 0   0 1 1

1 0 0   1 0 1   1 1 0   1 1 1

FIGURE 2.3 There are eight patterns that a string of three bits can exhibit.

Every bit that is added to a bit string doubles the number of patterns that the string can exhibit. This leads us to a very useful generalization to the question of how many patterns a bit string can exhibit. More specifically, we note that the number of patterns that a bit string of length N can exhibit is 2^N. Figure 2.4 gives insight into this pattern.

Real-world information can then be encoded as data by arbitrarily associating pieces of information with a particular bit pattern. We might, for example, associate the color red with the pattern 100, the color green with the pattern 010, and the color blue with the pattern 001 in a bit string of length three. As another example, consider encoding all of the symbols on a keyboard; including letters, digits, and punctuation symbols. We would make a list of every possible keyboard symbol and then begin to associate each symbol with a unique bit string pattern. We might, for

Length of Bit String	Number of Patterns
1	$2^1 = 2$
2	$2^2 = 4$
3	$2^3 = 8$
4	$2^4 = 16$
5	$2^5 = 32$
8	$2^8 = 256$
N	2^N

FIGURE 2.4 The number of patterns generated by bit strings.

example, associate the letter *A* with the 8 bit string 01000001 and a period with the 8 bit string 00101110.

2.3 DATA CAPACITY

Data encoding requires us to know how many bits are required to store a piece of information. When storing the symbols on a keyboard, for example, we must know how many bits would be required to store any one of the symbols. Would it be possible to encode a keyboard symbol as a 3 bit string or a 4 bit string rather than an 8 bit string?

Consider the following scenario. You are given a list of all keyboard symbols and you are also given a bit string of length three. Your task is to generate a unique bit pattern for each of the symbols in your list. You begin with the symbol "A" and decide to encode this symbol as 000. As you move through the alphabet, you choose to encode "B" as 001, "C" as 010, "D" as 011, "E" as 100, "F" as 101, "G" as 110, and "H" as 111. At this point you realize that there are no patterns left to encode the remaining symbols of the keyboard and hence a bit string of length three is not sufficient to encode one keyboard symbol.

In general, the number of bits required to store a piece of information is proportional to the number of values that the information may take. A single day of the week can be encoded as a bit string of length three since there are seven days in a week and there are eight patterns available. A single month of the year can be stored in a bit string of length four since there are 12 months in a year and 16 patterns available. Information that involves a large set of possible values will therefore require longer bit strings to encode while information of little content can be encoded in shorter bit strings. Figure 2.5 illustrates how the number of bits that are required to encode information of a certain type is related to the number of values associated with that type.

The *data capacity* of a computing system is the amount of information that can be encoded by the system. Since the data capacity is directly related to the number of bits that are available on the system, the data capacity is simply a count of the number of bits. Data capacity is not usually based on a direct count of the number of bits, but rather is based on the unit of measure known as a byte. A *byte* is a bit string of length eight. A single byte, therefore, is able to store 2^8 or 256 unique patterns.

One other measure of data capacity is known as a *word*, which is a unit of data capacity that is based on the hardware of a computing system. A word is a fixed-length sequence of bits that are processed as a single item

Type of Information	Number of Values	Number of Bits
coin toss	2	1
day of week	7	3
month of year	12	4
day of month	31	5
keyboard symbol	~104	7
day of year	365	9

FIGURE 2.5 The approximate number of bits required to store various types of information.

by the processor. The number of bits in a word varies by computing system but will typically be a multiple of eight. You may have heard of 32 bit processors or 64 bit systems. These phrases describe the word-length of a particular computing system. Common word lengths include 8, 16, 32, and 64.

Prefixes are used as multipliers to measure very large data capacities and the symbol B is used to denote a single byte. The computing industry uses terms such as kilobyte, megabyte, and gigabyte, and corresponding symbols KB, MB, and GB as common measures of data capacity. Figure 2.6 shows the most common prefixes and their meaning as both a power of 2 and an approximate decimal value.

Prefix	Symbol	Power of 2	Decimal
Kilo	K	2^{10}	$\sim 10^3$
Mega	M	2^{20}	$\sim 10^6$
Giga	G	2^{30}	$\sim 10^9$
Tera	T	2^{40}	$\sim 10^{12}$
Peta	P	2^{50}	$\sim 10^{15}$

FIGURE 2.6 Data capacity prefixes.

Type of Information	Data Capacity (Bytes)
keyboard symbol (letter)	1 B
10 page paper	40 KB
five minute MP3 audio recording	5 MB
high resolution digital picture	5 MB
CD audio disk	800 MB
DVD	8.5 GB
all of Wikipedia	6 TB*

* Wikipedia allows users to download the content of the entire web site. At the time of this writing, the content consisted of approximately 6 TB of data.

FIGURE 2.7   Amount of memory required to store certain types of information.

Computing systems are used to store vastly different types of information. Digital music players can record, store, and play back vast libraries of audio recordings. Cell phones can store and display high-quality video streams, while other systems gather and analyze massive amounts of scientific data for weather prediction or advancing scientific knowledge. The data capacity required by various types of information varies since the richness of the information content varies by type. Figure 2.7 shows how much data capacity is required to store certain types of information.

2.4 DATA TYPES AND DATA ENCODING

We have already asserted that all digital data is encoded as a sequence of bits and that information is associated with the various patterns that these bits may exhibit. This section gives further insight into how specific types of information are encoded and describes some of the inner workings of a computing system. We begin by discussing how numbers are encoded and then give a brief overview of how colors, pictures, and sound can be encoded.

2.4.1 Numbers

2.4.1.1 Numeral Systems

A *numeral system* is a way of representing numbers in written form. Consider, for example, the three numbers shown in Figure 2.8. If these markings are interpreted using the numeral system known as *tally*

FIGURE 2.8   Three numbers.

marking, they correspond to the numbers one, two, and three. Under tally marking, a number is represented by making one tally mark for each unit in the number. You may have used tally marking when keeping score in a game of tic-tac-toe or when counting the number of days you have spent in jail.

If these markings are interpreted using the *Roman numeral* system, they also correspond to the numbers one, two, and three. While the Roman numeral system has well-defined rules for representing integer numbers, the system is not widely used today because it is difficult to understand all of the rules and to decode larger numbers.

If these markings are interpreted using the *decimal numeral* system, they correspond to the numbers one, eleven, and one hundred and eleven. While the rules of the Roman numeral system are rather complicated, the rules of the decimal numbering system are simple enough to easily decode even large numbers. The decimal numeral system is a base 10 positional numbering system.

If the markings of Figure 2.8 are interpreted using the *binary numeral system*, they correspond to the numbers one, three, and seven. The binary system is very similar to decimal in that it is a positional numbering system such that any integer number is given by a sequence of digits where the position of each digit denotes a power of 2 and the value represents a multiplier. The only real difference between decimal and binary is that while decimal is a base ten numbering scheme, binary is a base two numbering scheme.

Of course these different numeral systems begin to look different once the numbers become larger. Figure 2.9 shows how the number five

✝✝✝✝✝ V 5 101

FIGURE 2.9   Four representations of the number five.

is represented in each of these four numeral systems: tally marking, the Roman numeral system, the decimal system, and the binary system.

In the tally marking system, all tally marks carry the same meaning: a value of one. The number represented by a particular tally is simply the sum of all the tally marks. In positional numbering systems, however, the digits of a number carry a meaning that is dependent upon where the digit is positioned within the number. Take the decimal number 323 for example. In this number, the first digit means three hundreds and the last digit means three ones.

2.4.1.2 Positional Numeral System

A *positional numeral system* must first specify a *base*, also known as the *radix*, that describes how many digits exist in that particular system. In the decimal numeral system, for example, the base is 10 since it uses the 10 digits 0 through 9. The smallest digit of a positional numeral system is always zero while the largest digit is always one less than the base. The decimal system is the most common system used by people, perhaps since we each have ten fingers that can be used to reason about numbers! But while the decimal system is the most commonly used system for human beings, the binary numeral system is the most commonly used in computing. The binary numeral system, as the name implies, uses only two digits. Digital electronics are well suited for binary systems since digital circuitry can be readily designed around numbers composed of only two digits. The standard positional numeral systems differ from one another only in the base they use.

Any positive integer greater than one can be used as a base in a positional numeral system. Since there are an infinite number of bases there are an infinite number of positional systems. Nonetheless, there are only a handful of commonly used numbering systems and these are listed in Figure 2.10. Notice that for numeral systems having a base greater than 10, the symbols used for digits will include glyphs other than those of the

Name	Base	Digits
Binary	2	0,1
Octal	8	0,1,2,3,4,5,6,7
Decimal	10	0,1,2,3,4,5,6,7,8,9
Hexadecimal	16	0,1,2,3,4,5,6,7,8,9,A,B,C,D,E,F

FIGURE 2.10 Common positional numbering systems.

$$\text{Number} \longrightarrow \quad 9 \quad\quad 2 \quad\quad 5 \quad\quad \text{Base} \longrightarrow 10$$

$$\text{Position} \longrightarrow \quad 2 \quad\quad 1 \quad\quad 0$$

$$\text{Power of Base} \longrightarrow \quad 10^2 \quad 10^1 \quad 10^0$$

$$\text{Meaning} \longrightarrow \quad 9 \times 10^2 + 2 \times 10^1 + 5 \times 10^0$$

FIGURE 2.11 Example of using the decimal numeral system.

decimal system. As an example consider the hexadecimal system, which uses the symbol A to denote the value 10, the symbol B to denote the value 11, and onward through F to denote the value 15.

In a positional number system, a value is represented by a sequence of digits where the position of each digit in the sequence is associated with an integral power of the base and the digit is a multiplier for that base term. The position of each digit is numbered starting from zero on the right and increasing as the digits move right to left. The right-most digit is therefore always at position 0, the next digit moving left is at position 1 and so forth.

Consider the example of Figure 2.11 that shows how to interpret the decimal number 925. For this number we see that the digit 5 is at position 0, the digit 2 is at position 1 and the digit 9 is at position 2. The radix is assumed to be 10 and hence the positional terms will be integral powers of 10. We then understand that the digit 5 is a multiplier for 10^0, the digit 2 is a multiplier for the 10^1, and the digit 9 is a multiplier for 10^2. In other words, the number 925 is understood as $(9 \times 10^2) + (2 \times 10^1) + (5 \times 10^0)$ and since any positive integer that is raised to the 0th power is equal to 1 this reduces to $(9 \times 100) + (2 \times 10) + (5 \times 1)$.

The most common powers of ten have names that allow us to more easily describe decimal numbers. These names are shown in Figure 2.12. Note

Name	Value
Ten	10^1
Hundred	10^2
Thousand	10^3
Million	10^6
Billion	10^9
Trillion	10^{12}
Googol	10^{100}

FIGURE 2.12 Common names given to various powers of 10.

$$Number \rightarrow 1 \quad 0 \quad 1 \quad Base \rightarrow 10$$
$$Position \rightarrow 2 \quad 1 \quad 0$$
$$Power\ of\ Base \rightarrow 10^2 \quad 10^1 \quad 10^0$$
$$Meaning \rightarrow 1\times10^2 + 0\times10^1 + 1\times10^0$$

FIGURE 2.13   Interpretation of 101_{10}.

that a googol corresponds to the 100th power of 10 and is the namesake for the Internet search engine powered by Google.

People almost always use the decimal numbering system when writing numbers and therefore usually assume that the base is 10. In situations where the radix may not be clear, however, it is useful to express the base by using a subscript notation. For example, the subscript in 101_{10} indicates that the number is expressed in the decimal system, whereas 101_{16} is a number that is expressed using a base of 16 and 101_2 is a number that is expressed in binary. Be careful to note that these numbers are not different ways of writing the same number but are different numbers! Because computational thinkers often use numbers that are represented in different base systems, we will use subscript notation to make the radix clear. Whenever we write a number without a subscript we assume that it is expressed in the decimal numeral system.

Consider, for example, the two numbers 101_{10} and 101_2 as shown in Figures 2.13 and 2.14, respectively. The number 101_{10} has the value one hundred and one while the binary number 101_2 does not.

Since the radix is 2 we understand this to mean $(1 \times 2^2) + (0 \times 2^1) + (1 \times 2^0)$ which is equal, in the decimal system, to $4 + 0 + 1$, or 5.

2.4.1.3 Integers as Binary Bit Strings

Computing systems represent integer numbers as binary bit strings. The binary system is a fitting choice for computers because there are only two values in the binary system and hence a single bit has sufficient data capacity to store a single binary digit. If we assume that the patterns of the bit

$$Number \rightarrow 1 \quad 0 \quad 1 \quad Base \rightarrow 2$$
$$Position \rightarrow 2 \quad 1 \quad 0$$
$$Power\ of\ Base \rightarrow 2^2 \quad 2^1 \quad 2^0$$
$$Meaning \rightarrow 1\times2^2 + 0\times2^1 + 1\times2^0$$

FIGURE 2.14   Interpretation of 101_2, which is equal to 5_{10}.

Bit String	Decimal Value
00000000	0
00000001	1
00000010	2
00000011	3
00000100	4
...	...
11111110	254
11111111	255

FIGURE 2.15 Understanding binary bit strings.

string correspond directly with how the number is written in the binary (or base 2) numeral system, it is straightforward to see how an integer is stored. Figure 2.15 shows the bit patterns and corresponding numeric values for a select number of numeric values.

An 8 bit string can only represents 256 numbers since an 8 bit string can only exhibit 2^8 unique patterns. This implies that the largest number that can be encoded as an 8 bit binary number is $2^8 - 1$ or 255. This statement can be generalized by stating that any binary bit string of length N can only encode the numbers 0 through $2^N - 1$.

2.4.1.4 Real Numbers as Binary Bit Strings

Although we can represent an integer as a sequence of bits, is it possible to represent a real number such as 2.31 or 2.125 or even 3.1415926? Consider extending the meaning of a positional numbering system to the right of the decimal such that the positions start at –1 and decrease with distance from the decimal. For example, in base 10, the value 1.625 means (1×10^0) + (6×10^{-1}) + (2×10^{-2}) + (5×10^{-3}). We can then represent real numbers as binary bit strings assuming that we can determine the decimal location. Consider, for example, the meaning of 1.101_2.

$$1 = 101_2 = 1 \times 2^0 + 1 \times 2^{-1} + 0 \times 2^{-2} + 1 \times 2^{-3}$$
$$= 1 \times 1 + 1 \times \frac{1}{2} + 0 \times \frac{1}{4} + 1 \times \frac{1}{8}$$
$$= 1 + \frac{1}{2} + \frac{1}{8}$$
$$= 1.625$$

2.4.1.5 Precision as a Source of Error

One difficulty that arises when encoding real numbers is that there may be an arbitrary number of digits on the right side of the decimal value to represent even small numbers. If we write 1/3 as a real number, for example, we begin to write 0.333333333333333333333333 but soon realize that we will never be able to write as many digits as are required for a completely accurate representation. Most real numbers cannot be exactly encoded by the encoding scheme described earlier and are therefore represented as an approximate value. Computer applications typically, therefore, allow rounding errors to occur and hence computers often produce incorrect, although highly accurate, results. *Precision* is a measure of the accuracy of a stored quantity. The precision is usually measured as the number of available bits. If a computer uses 16 bits to store real numbers, we say that the computer is precise up to 16 bits, or that the computer uses 16 bit precision.

2.4.1.6 Underflow and Overflow as Sources of Error

Other kinds of error can be introduced into a computing system through *underflow* and *overflow*. Recall that an 8 bit binary string can only hold values between 0 and 255. An example of overflow occurs when a computer is instructed to add 1 to the value 255 and store the result as an 8 bit binary string. Of course the result should be 256 but since 256 cannot be encoded as an 8 bit binary string the result will either be an error of some sort or, on many computing systems, the result will actually wraparound to 0! In general, overflow occurs when a computer instruction produces a value that is too large to be encoded by the number of bits available. Underflow occurs when a computer instruction produces a value that is too small in magnitude (i.e., very close to zero) to be encoded by the number of bits available.

2.4.2 Text

All data that is stored in a computing system is encoded as bit strings. You might wonder, then, how can a bunch of bits be made to look like text? Even the words that you are now reading are stored in binary form and this information is then presented to you as text. The text you are reading even has different fonts using different sizes and weights.

First, it is important to understand that a textual character, when drawn on either a page of text or on a computer screen, is really just a picture. Figure 2.16 illustrates the graphical nature of text by showing the letter Q as it appears in different font families. Although the pictures that are used

FIGURE 2.16   The letter Q is shown using five different font families.

to represent the letter Q are each different, they nonetheless each represent the same thing: the letter Q.

Textual characters are usually encoded as integer values using the encoding schemes discussed earlier in this chapter. Each number is arbitrarily associated with the image that should be used when the character is drawn on a page or shown on the computer screen. The associations between numbers and text are known collectively as a character encoding scheme. The most common character encoding scheme is the ASCII table, shown in Figure 2.17, which defines associations between numbers and English textual characters. In the ASCII table, the number 65 is associated with uppercase A, whereas the number 97 is associated with lowercase a. The number 38 is associated with the ampersand (&) and the number 126 is associated with the tilde (~). Since English has relatively few characters, the *ASCII table* only has about 128 entries. Other languages, such as Japanese or Chinese, use thousands of characters and hence the association tables are much larger.

Some text contained in the ASCII table is not really pictorial but rather a command that a text editor or text processor must follow. These characters are known as nonprintable text characters because they cannot be drawn. The backspace key, for example, is not printable since you do not see anything when you strike that key, but you would expect that a text processor would take some action whenever the user enters a backspace. Nonprintable characters are shown in Figure 2.17 as an empty cell.

In addition, the images associated with the textual characters are dependent on the specific font or typeface. Many fonts use the same basic symbols, but draw the glyphs in different ways (serif, sans serif, boldface, italics, Helvetica, Times, etc.). Applications that process text must usually associate a font with a textual character but this association is beyond the scope of our discussion.

2.4.3 Colors

The human eye perceives color through three types of biological photosensors known as *cones*. Each cone is attuned to one of three wavelengths that correspond roughly to red, green, and blue light. The individual responses

0	1	2	3	4	5	6	7	8	9	10	11	12	13	14	15	16	17	18	19
													!	"	#	$	%	&	'
20	21	22	23	24	25	26	27	28	29	30	31	32	33	34	35	36	37	38	39
(	)	*	+	,	-	.	/	0	1	2	3	4	5	6	7	8	9	:	;
40	41	42	43	44	45	46	47	48	49	50	51	52	53	54	55	56	57	58	59
<	=	>	?	@	A	B	C	D	E	F	G	H	I	J	K	L	M	N	O
60	61	62	63	64	65	66	67	68	69	70	71	72	73	74	75	76	77	78	79
P	Q	R	S	T	U	V	W	X	Y	Z	[	\	]	^	—	`	a	b	c
80	81	82	83	84	85	86	87	88	89	90	91	92	93	94	95	96	97	98	99
d	e	f	g	h	i	j	k	l	m	n	o	p	q	r	s	t	u	v	w
100	101	102	103	104	105	106	107	108	109	110	111	112	113	114	115	116	117	118	119
x	y	z	{	|	}	~													
120	121	122	123	124	125	126	127	128	129	130	131	132	133	134	135	136	137	138	139

FIGURE 2.17 The ASCII table associations between numbers and textual characters.

of all cones combine to form the perception of a single color at a single point within the field of view. The design of this biological system suggests that color is a three-dimensional entity.

The *RGB* color model is the most common way of representing color in image-processing systems. The RGB model uses red (R), green (G), and blue (B) as the primary colors such that any color can be created by combining different amounts of these three primaries. By way of example, consider a flashlight that has a slider allowing you to choose the strength of light emitted. In setting the slider to 0, the flashlight is turned completely off and generates no light, whereas in setting the slider to 255 (the maximum setting) the flashlight generates as much light as it is capable of generating. Now consider three such flashlights: the first emits purely red light, the second emits purely green light, and the third emits purely blue light. If all three flashlights are aimed at the same spot on a white wall any color can be projected onto the wall by adjusting the slider values on the three lights in different ways. If all sliders are set to 0, black is projected onto the wall. If all sliders are set to 255, white is projected onto the wall, and if all sliders are set to 128 then gray is projected.

In computing systems, a *color* is usually encoded as three integer numbers where each number is in the interval 0 to 255. In addition, since each value can be one of only 256 different values, a bit string of length eight is

Color	Name	Bits			Decimal
		red	green	blue	
	red	111111110000000000000000			(255,0,0)
	green	000000001111111100000000			(0,255,0)
	yellow	111111111111111100000000			(255,255,0)

FIGURE 2.18 How colors are typically encoded as binary bit strings.

sufficient for encoding each value. This implies that a single color will be encoded as a bit string of length 24 since there are three values and each value is encoded with 8 bits. Figure 2.18 shows how the three colors red, green, and yellow (which is a mixture of red and green) would typically be encoded as 24 bits within a computing system. The corresponding interpretation of the three decimal values is given in the right-most column.

2.4.4 Pictures

The most common encoding of a digital image is that of a two-dimensional grid of colors. Each element of the grid is known as a *pixel* (this term is an abbreviation of the phrase "picture element") and represents a very small rectangular region of the image that is comprised of a single color. When the grid of pixels is viewed at an appropriate distance, the individual pixels recede from view while collectively appearing as a naturally colored and smooth image.

Images are typically encoded as a sequence of pixels where each pixel corresponds to a single color. Since a color is normally encoded as a 24 bit string, the total bits required to encode a digital image is on the order of 24 bits times the number of pixels in the image. A small number of extra bits are also required to store useful information such as the width and the height of the image. This extra information is referred to as a header, and because the header can be encoded using an extremely small number of bits relative to the pixel data, we will not consider it further.

High-definition video uses individual images that have up to 1920 columns and 1080 rows of pixels for a total of 1920 × 1080 or 2,073,000 pixels. Since each of these pixels is encoded as a bit string of length 24, the image can be encoded as a bit string of length 1920 × 1080 × 24 or 49,766,400 bits. Alternatively, using a byte as a measure of data capacity, the image can be encoded using 1920 × 1080 × 3 or 6,220,800 bytes or approximately 6 megabytes. Digital cameras take images that use on the

0	0	1	1	1	1	0	0
0	1	0	0	0	0	1	0
1	0	1	0	0	1	0	1
1	0	0	0	0	0	0	1
1	0	1	0	0	1	0	1
1	0	0	1	1	0	0	1
0	1	0	0	0	0	1	0
0	0	1	1	1	1	0	0

FIGURE 2.19 A black-and-white image could be encoded using the bits shown on the right.

order of 10 megapixels and images of this resolution can be encoded as a string of 30 megabytes.

Sometimes a picture does not contain many colors. Black-and-white images, for example, use only two colors, whereas grayscale images use at most 256 colors; each color being a unique achromatic gray. While a color image generally uses 24 bits per color, other image types may use fewer than 24 bits per color since there are many fewer possible color values. Since grayscale images may use no more than 256 colors, we can encode a grayscale color using 8 bits and since there are only two colors in a black-and-white image, we can encode the colors using a single bit!

Figure 2.19 shows a black-and-white image having eight columns and eight rows for a total of 64 pixels. Since there are only two colors, we use a single bit to encode a color. In this figure we use 0 to denote white and 1 to denote black. When these bits are used to encode the image the bits are arranged as a single linear unit in memory and not organized as the two-dimensional structure illustrated in Figure 2.8. More specifically, the image pixels could be encoded as the bit string 00111100010000101010010 1100000011010010110011001010000100011110 0.

2.4.5 Sound

Digital audio is used almost exclusively today when recording, processing, and distributing sound. Online music stores contain hundreds of thousands of hours of digital sound; the soundtracks and audio of DVD movies are encoded digitally; and your voice is digitally encoded by your cell phone prior to being delivered to the person at the other end of the call.

Sound is a physical phenomenon caused by waveforms that propagate through the air. A microphone is used to transform the waveform into an analog electric signal after which the analog signal is sampled to produce a digital encoding. Sampling is a process where the strength of a changing signal is measured at regular time intervals and those measurements are then recorded. In this way, the sound wave is converted into a sequence of numeric values.

Frequency is the rate at which sound waves change and is measured in terms of the hertz. The *hertz* is denoted as Hz and is defined as the number of cycles (or changes) per second. Sound waves that change at a slow rate are perceived as a low pitch while sound waves that change at a high rate are perceived as higher in pitch. The average person can hear sound waves 20 Hz up through about 20,000 Hz or 20 kHz.

Figure 2.20 shows how the strength of a sound wave is sampled at regular periods of time to obtain a sequence of samples. In this example, the sound wave has a frequency of approximately 4 Hz (although this is not obvious) and must therefore be sampled at least eight times per second to adequately record the signal. The actual sampling rate is 10 Hz, or ten times per second. Every tenth of a second the strength of the sound wave is measured and recorded. The resulting digital sound is depicted as the sequence of measurements shown on the right of Figure 2.20. It should be apparent that higher sampling rates will yield more accurate encodings of the analog signal.

The *Nyquist–Shannon sampling theorem,* named after Harry Nyquist and Claude Shannon, states that the rate at which samples are taken must be at least twice that of the highest frequency signal that is to be measured. This theorem implies that a sampling rate of at least 40 kHz is required when capturing sound waves that span the entire 20 to 20,000 Hz range of human hearing. This fact explains why compact discs (CDs) are sampled

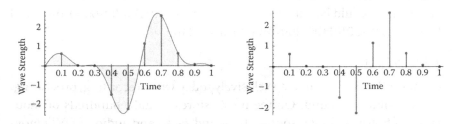

FIGURE 2.20 A sound wave is sampled ten times per second to obtain the sequence plotted on the right.

at a rate of 44.1 kHz, digital audio tapes (DATs) are sampled at 48 kHz, and a 96 kHz sampling rate is used for digital video discs (DVDs).

Consider a song that is five minutes long and is sampled at a rate of 48 kHz for recoding on a digital audio tape. To determine the number of samples required to encode the song we must note that there are 48,000 samples taken every second and that the song lasts for 300 seconds. Thus, the song will be encoded as a sequence of $48000 \times 5 \times 60$ or 14,400,000 samples. For this example, we will assume that a single sample is encoded as an 8 bit string, which means that the song can then be encoded in 14,400,000 bytes or about 14 MB. Of course if the number of bits required to encode a single sample is greater than 8, the number of bytes required to encode the song also changes. Compact discs, for example, use 16 bits for each sample and DVD mastering usually requires 24 bits per sample.

2.5 DATA COMPRESSION

The previous section described techniques for converting real-world information into digital data and did not make any attempt to minimize the number of bits used for encoding. *Data compression* is a technique that is used to encode real-world information using fewer bits than a straightforward encoding requires. In this section we will briefly describe why data compression is useful and how data compression can reduce the number of bits required to encode information.

Data compression is often very useful because it reduces the consumption of resources such as hard disk space or the time it takes to send files across the Internet. If a digital image can be compressed from 10 MB in size to 1 MB in size, for example, the compressed image can be sent across the Internet ten times more quickly than the uncompressed image. Also, if a photo album that consumes 200 GB of image data can be compressed to 20 GB, then you may not need to buy a new hard drive to store your photo album!

There is a cost to using data compression, however, since a great deal of computational work is required for a computer to take the uncompressed file and find a way to eliminate some of the bits. Also, when a file is stored in compressed form, the computer must take time to uncompress the file for further processing. This computational work also takes time to perform and sometimes the cost of compression outweighs the benefits. Video compression, for example, takes a great deal of time to perform and may even require specialized hardware to perform fast enough for live streaming.

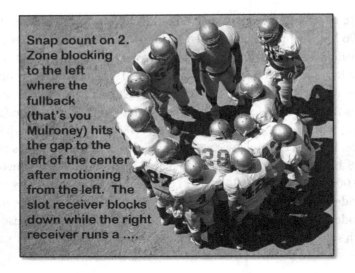

Snap count on 2. Zone blocking to the left where the fullback (that's you Mulroney) hits the gap to the left of the center after motioning from the left. The slot receiver blocks down while the right receiver runs a ....

By way of example, consider the reasonably complicated sport of American football. Two teams of eleven players compete to move the football downfield into the opponent's end zone. The game consists of a series of individual plays where every player on a team must know what to do on the play. Teams will often try to gain an advantage by executing plays more quickly than their opponent can respond. The problem is that a single play may take many words to fully convey since every play must describe things like the blocking scheme, the snap count, the position of the running backs, and the direction and nature of the play itself. A full play might, for example, be described as "The snap count is 3. We will use a zone-blocking scheme to the right. The fullback must line up to the right, the halfback will motion to the left, and the quarterback will hand off to the right." Calling such a play by fully describing it with full sentences is simply too time consuming for effective game play. As a result, teams have developed ways to communicate the same information by using many fewer words. A simple play-calling scheme might use four digits to communicate a single play. The first digit might indicate the snap count, the second the blocking scheme, the third will define the alignment and motion of the backs, and the fourth will describe the nature of the play itself. The play 3518, for example, might be understood as 3-snap count; 5-zone blocking to the right; 1-fullback aligns right, halfback motions left; 8-quarterback hands off to the right. This compression technique takes work by the players to memorize and learn what the numbers actually mean but this cost occurs before the game is played. The payoff comes during the game when the play calling can be quickly communicated (transmitted) and decoded.

The obvious and central idea in data compression is that unnecessary bits should be eliminated from the encoding scheme. While this is an obvious idea, identification and elimination of unnecessary bits is often a very difficult task and largely depends on the type of information being compressed: text, images, or audio.

2.5.1 Run-Length Encoding

Run-length encoding is one of the simplest types of compression techniques that can be used on images and even text. We will describe how run-length encoding can be used to encode multiple pixels as a single value rather than recording each pixel individually. Consider, for example, a binary image in which each pixel is encoded as a single bit that is either 0 (denoting black) or 1 (denoting white). One row of the image may contain the 32 pixel sequence

$$11111111110001111111111111111111$$

This row contains three runs: 10 white pixels followed by 3 black followed by 19 white. This information can be encoded as the 3 byte sequence {10, 3, 19} by assuming that the data begins with white runs. Note that the raw representation uses 32 bits of memory while the run-length encoding uses only 24 bits of memory if we assume that 8 bit bytes are used to store each run.

Figure 2.21 gives a complete example of how run-length encoding can be used to compress binary image data. The heart image has 18 columns and 14 rows. Raw encoding, where a single bit is used to encode each pixel, requires 18×14 or 252 bits. Run-length encoding, where we encode each row by recording the lengths of each run in the row, uses significantly fewer bits. The first row, for example, is encoded by the five numbers {3, 3, 5, 4, 3} since the first row starts with three white pixels followed by three black pixels and so on. If we count the total number of runs in the entire image, we find that there are a total of 43 runs. Since the longest run is 18, we realize that we can encode each run using five bits. Therefore, the total number of bits required to run-length encode this image is given as 43 runs times 5 bits per run for a total of 43×5 or 215 bits.

You may notice that some runs, rows 3 to 7 for example, begin with runs of length zero. This is because when we decompress the run lengths

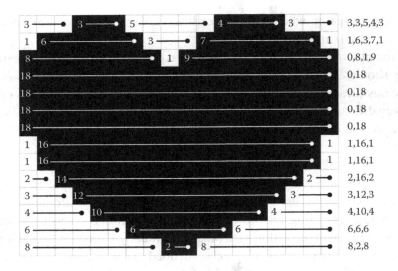

FIGURE 2.21 A black-and-white image could be encoded using the numbers shown on the right.

we must assume that we are starting with either white or black runs. In this example, the assumption is that each row starts with a run of white pixels. Since there are no white pixels at the beginning of these rows, we encode the runs as being length zero. When decompressing the run-length data, we must also know how many columns and rows are in the image. These two pieces of information must be stored in the image header.

Compression schemes can be classified as either *lossless* or *lossy*. A lossless compression scheme encodes an exact representation of the original data, whereas a lossy compression scheme encodes an approximate representation. The central idea behind lossy compression is to identify pieces of information that are of little importance and simply discard those pieces of information in order to use fewer bits. Such compression schemes are known as lossy because real information is actually lost in the process.

There are hundreds of image file types in common use. Among the most popular are the PNG, JPEG, and GIF formats that are most often used to display images on the Internet. Image file types typically use some compression scheme to reduce the overall size of the image file. The PNG image format uses lossless compression to reduce the overall file size, whereas JPEG and GIF typically use lossy compression. While the JPEG and GIF file formats do include limited support for lossless compression, almost all of the JPEG and GIF images on the Internet have been compressed using a lossy technique.

| Uncompressed image:
8,224,416 bits | JPEG
(Lossy): 55384 bits | PNG
(Lossless): 5,668,304 bits |

FIGURE 2.22   Lossy and lossless image compression.

Figure 2.22, a digitized image of a famous painting by James McNeill Whistler [1], illustrates the difference between lossy and lossless compression. The digital image consists of 513 columns and 668 rows and is stored in uncompressed form using 24 bits per pixel. The uncompressed image therefore requires 513 × 668 × 24 or 8,224,416 bits. JPEG compression reduces the memory requirement to 55384 bits but significantly reduces the image quality. PNG compression reduces the memory requirements to 5,668,304 bits without any reduction in quality. JPEG can be controlled to provide much higher quality results but with a corresponding loss of compression effectiveness.

Sound information can also be compressed by recognizing that certain sound waves are less important than others when listening to music or voice recordings. The MP3 standard, for example, will discard higher frequency sound waves (or portions thereof) in order to encode the sound using fewer bits.

2.6 SUMMARY

If computers are used to process information about the real world, this information must first be converted into a computable form. This chapter has given an overview of how real-world information of various types: numbers, text, images, and sound can be converted into computable form. Bit strings are used to encode all computable data. The length of a bit string gives a measure of the amount of information that the bit string can hold such that each unique pattern generated by the bit string corresponds to some piece of real-world information.

REFERENCE

1. Wikipedia, http://en.wikipedia.org/wiki/Symphony_in_White,_No._1:_The_White_Girl (accessed January 2013).

TERMINOLOGY

analog	information
ASCII table	lossless
base	lossy
binary digit	numeral system
binary numeral	Nyquist-Shannon sampling
bit	theorem
bit string	overflow
byte	pixel
color	positional numeral system
cones	precision
continuous	radix
data	RGB
data capacity	Roman numeral
data compression	run-length encoding
decimal numeral	sound
digital	tally marking
discrete	underflow
frequency	word
hertz	

EXERCISES

1. Take the initials of your first and last name and write them down. These are initials, so they should be capitalized! Write down the 16 bits that a computer would most likely use to store these two initials in a computer. (Hint: The encoding is given by the ASCII table.)

2. What text is encoded in the following? (Hint: The encoding is given by the ASCII table.)

   a. 00110010001010110011001100111101001 10101

   b. 0100001001101001011101000 1110011

   c. 0011110000110011001100 11

3. Write the number 4 using the following systems.

   a. tally mark number system

   b. Roman number system

   c. binary system

4. Write the equivalent decimal (base 10) number for each of the following binary numbers.

   a. 10011

   b. 11000

   c. 10010

   d. 00000

   e. 11111

5. For each quantity listed, indicate whether the quantity is continuous or discrete.

   a. Day of the week

   b. Flying speed of a hummingbird

   c. Number of peas in a pea pod

   d. Length of a commencement speech

   e. Number of chapters in a book

    f. Number of commas in a book

    g. Weight of a book

    h. Age of a book

6. For each pair of the following items, circle the one that would most probably require more bits to represent.

    a. The book *Peter Rabbit* or a single JPEG image from a modern digital camera

    b. One pop song or one page of the textbook

    c. One color image saved as a JPEG from your digital camera or the complete works of William Shakespeare (Make sure to give the average size of a JPEG image from your camera OR assume that each JPEG image consumes 1.5M bytes.)

7. How many unique patterns does a sequence of 5 bits generate?

8. How many unique patterns does a sequence of N bits generate?

9. For each quantity listed, give the number of bits that would be required to store the data in a computing system.

    a. The month

    b. A class grade. Class grades must be one of A, B, C, D, F, or I (for incomplete).

    c. An MPAA movie rating. The MPAA rates movies as one of the following: G, PG, PG-13, R, NC-17.

    d. The track number of a song on an MP3 album. The most tracks that can be put on an MP3 album is 100.

    e. The day of the year. There are 365 days in a year (except for leap years in which there are 366 days). For example, January 1 would always be day 1 while December 31 would be day 365 for non-leap years and day 366 for leap years.

    f. One nucleobase of a DNA string. There are only four nucleobases in this example: adenine (A), guanine (G), cytosine (C), and thymine (T).

10. Consider a song that is two minutes long and is sampled at a rate of 24 kHz for recoding on a digital audiotape. We will assume that a sample is encoded as an 8 bit string. How many bits are required to store this song without compression?

11. Consider a black-and-white picture that consists of 9 columns and 12 rows of pixels. The picture is encoded using the following run-length encoding where a forward slash (/) denotes the end of a row. Draw the picture.

    0,8/1,3,4/2,2/2,2/2,2,3,1/2,6/2,2,3,1/2,2/2,2/2,2/2,2/1,4

12. Consider the following black-and-white picture. Give the run-length encoding of the picture.

CHAPTER 3

Logic

Logic is the anatomy of thought.

—**JOHN LOCKE**

OBJECTIVES

- To understand that logic is necessary and useful for correct and rational thought
- To describe how the logic of natural language is expressed symbolically
- To define logical values and the logical operators AND, OR, NOT, IMPLIES
- To define truth tables, tautologies, and contradictions
- To describe how logic can be applied to solve a variety of real-world problems

---

Your life is full of daily decisions that, over time, determine the trajectory of your personal history. Many of these decisions involve making significant choices about who to marry, what career path to undertake, what to eat, how to best manage your finances, or what political cause to support. Every decision that you make in your life is based on some belief about the nature of the universe and, perhaps, the very meaning of life itself. It is apparent if we adopt beliefs that are false, then decisions that are based on those beliefs will be misguided at best and destructive at the worst. Even if our belief system includes only true beliefs, we might still make decisions through illogical or haphazard reasoning about those beliefs and again, end up making misguided or potentially destructive choices.

In this chapter we will see that logic serves as the basis for all correct and rational thought. We will discover that certain patterns of thinking will form a path of truth, whereas other patterns will inevitably lead to falsehood. We will also discover that these patterns of thought can be expressed in ways that a computer can understand and process. Finally, we will show that these patterns of thought are pervasive and can be used to solve a wide variety of real-world problems.

3.1 WHAT IS LOGIC?

You may be surprised to hear that numerous mathematicians and philosophers have defined human reasoning as a logical system. *Logic*, in its broadest sense, deals with correct and incorrect ways to reason. Logic provides a way to tell the difference between incorrect and correct thinking, and can therefore be defined as the *science of correct thinking*. The study of logic can be broken into two categories: inductive logic and deductive logic.

Inductive logic is a type of reasoning that begins with a set of observations or experiences from which conclusions can be derived with some degree of certainty. As an example of inductive reasoning, consider a scenario where you have eaten brussels sprouts only twice in your life. Unfortunately, you became physically ill within hours of eating them and therefore conclude you are allergic to brussels sprouts. Although the conclusion that "I am allergic to brussels sprouts" is reasonable, it is not the only conclusion that could be reasonably reached and is therefore not certain. Perhaps you were coincidently exposed to a contagious virus shortly before eating the brussels sprouts and this virus was the actual cause of your illness. Conclusions reached in an inductive system are necessarily uncertain and are directly related to the number of experiences on which the conclusion is based. For example, the certainty of the conclusion that "I am allergic to brussels sprouts" would be increased if you ate them three more times and became physically ill in each case.

Deductive logic, by contrast, begins by assuming that certain things are absolutely true from which other facts must also be absolutely true. Any conclusions that are reached under deductive logic are absolutely, irrefutably, and certainly true if the assumptions are true. You may, for example, assume that all men are mortal and that Aristotle is a man. From these two assumptions you may deduce with absolute certainty that Aristotle is mortal.

Aristotle was an ancient Greek philosopher who lived around 350 BC. Aristotle described the logical rules of deduction that must be followed in order to draw correct conclusions from a set of assumptions. Aristotelian logic is a deductive system that uses human language to form *premises* from which a *conclusion* can be deduced if certain rules of logic are followed. A premise is a statement that is assumed to be true and that is used to justify a conclusion. A *conclusion* is a statement of truth that must logically flow from the premises. More specifically, Aristotle described a *syllogism* as a logical argument that contained two premises and a true conclusion that must logically follow if the two premises are actually true. As an example, consider the following syllogism.

- Premise: Computational thinking is beneficial for all students.

- Premise: Jemimah Farnsworth is a student.

- Conclusion: Computational thinking is beneficial for Jemimah Farnsworth.

Symbolic logic is a modern extension of Aristotelian logic where symbols, rather than phrases drawn from human language, are used to represent statements of truth. Logical operators, also known as logical connectives, are used to express logical thought. *Boolean logic* is a symbolic logic system that was created by a mathematician named George Boole around 1850. Modern computing systems rely heavily on the rules of Boolean logic for generating correct and reliable results. The following sections describe the elements of Boolean logic and show how Boolean logic can be understood and processed by computing systems.

3.2 BOOLEAN LOGIC

Propositions form the basic units of Boolean logic. A *proposition* is a statement that can be either true or false. Stated another way, a proposition is a statement that has a value of either true or false. Consider, for example, the following true statement.

"Mount Everest is the tallest mountain in the world."

Since it makes sense to say that it is true that Mount Everest is the tallest mountain in the world we know that this statement is a proposition because it is an expression of truth. It is very important to understand that although some propositions have a value that is true, propositions can also have a value that is false. For example, consider the following statement.

"The Mississippi is the longest river in the world."

The Nile is the longest river in the world and hence the statement that the Mississippi is the longest river in the world is a false statement. Just because the sentence is false, however, does not mean that the sentence is not a proposition. In fact, just the opposition conclusion holds. The sentence is shown to be a proposition precisely because it has a value of false. In other words, the value of this sentence is false.

Some sentences do not have a truth value and are therefore not propositions at all. Consider, for example, the following sentences.

"Brush your teeth."

"How tall is Mount Everest?"

The first sentence is a command, written in the imperative voice, and therefore has no truth value. We could not meaningfully say, for example, that "Brush your teeth" is true. The second sentence is a question and therefore has no truth value. Again, we cannot meaningfully say that "How tall is Mount Everest?" is false.

In Boolean logic there are only two values: true and false. These two values are known as *logical values* or *truth values*. Boolean logic does not allow for uncertainties or for probabilities. A proposition is either completely true or it is completely false. A proposition cannot be probably true or probably false. Nor can a proposition be both true and false at the same time. The notion that there are only two truth values in a logical system is referred to as the *law of the excluded middle*.

Propositions can be either *simple* or *compound*. A *simple proposition* is one that cannot be broken into parts. A *compound proposition* is formed by combining simple propositions with logical connectives, also known as logical operators. There are four fundamental logical operators that are best described by the words *and, or, implies,* and *not.* Consider, for example, the following two simple propositions that are connected using the *and* operator to form a compound proposition.

Simple: "I am hungry."

Simple: "I am cold."

Compound: "I am hungry and I am cold."

This compound proposition is a statement that has a truth value since the statement that "I am hungry and I am cold" is either true or it is false. Boolean logic allows us to concisely and precisely express lengthy propositions by (1) abbreviating simple propositions and (2) precisely defining each of the four logical connectives.

Simple propositions are often abbreviated, or labeled, in order to make logical expressions more concise and readable. The labels serve as an abbreviation for the entire proposition and are always given as a single capital letter to make things simpler. We might, for example, use the letter P as an abbreviation for the proposition "I am hungry." Throughout our text we will use an equal sign (=) to indicate the abbreviation that is being used for a particular proposition. Following are examples of abbreviating propositions.

P = "I am hungry."

Q = "I am cold."

S = "Mount Everest is the tallest mountain in the world."

We can then create new propositions by using the logical operators to connect existing propositions. Given the aforementioned abbreviations shown, we can create the proposition "I am hungry and I am cold" as shown by the following.

P and Q

3.2.1 Writing Well-Formed Propositions

Propositions must be *well formed* (i.e., properly written) to have meaning. When writing a proposition, certain rules must be exactly followed in order to construct a meaningful proposition. If these rules are not followed you may write something that appears to be a proposition but is actually just a meaningless jumble of labels and operators.

Perhaps you are familiar with the famous poem "Jabberwocky" written by Lewis Carroll. The poem looks and sounds like an English poem, but is in reality just a series of nonsensical syllables (when written) and melodic sounds (when spoken). The opening stanza is reproduced next.

> Twas bryllyg, and ye slythy toves*
> Did gyre and gymble in ye wabe:
> All mimsy were ye borogoves;
> And ye mome raths outgrabe.

Although "Jabberwocky" appears to be a meaningful poem, it is nonsense. There is no real meaning, because the phrases are not composed of well-defined words and the normal rules of grammar are not followed. Just as a poem must follow rules of grammar and spelling, a well-formed logical proposition must also follow specific rules. These rules guarantee that the proposition has meaning and is not merely a jumble of nonsense. The grammatical rules for writing a well-formed proposition are listed next.

Rule 1—Each of the following is a simple proposition.

a. Any single letter

b. True

c. False

---

* The slithy tove image is derived from an illustration by John Tennel in Carroll's original publication.

Rule 2—Let a box (□) stand for a proposition (either simple or compound). Assuming that each box is some proposition, then each of the following are also propositions.

    a. □ and □

    b. □ or □

    c. □ implies □

    d. □ ≡ □

    e. not □

    f. (□)

These rules establish patterns of grammatical structure that all propositions must possess. The various parts of Rule 2 specifically show how to make compound propositions by using logical operators to combine simpler propositions into a larger whole. With these rules in mind, consider the mixture of letters, operators, and parenthesis listed below.

P and not (Q or R)

We might naturally ask if "P and not (Q or R)" is a well-formed proposition (and therefore a meaningful proposition) or if it is merely a nonsensical jumble of letters and words. For this example we can prove that the expression is well-formed by carefully applying the two grammatical rules that well-formed propositions must follow. Our process of reasoning will move from the simplest substructures within the entire proposition up to the more complex propositions that comprise the whole.

Step 1—We can first apply Rule 1a to the capital letters P, Q, and R. In our proof, we will use a box symbol, □, to denote a portion of the text that has been proven to be a proposition by applying one of the rules. To prove that the entire expression is a well-formed proposition, we must show that the entire expression can be reduced to a single box.

Step 2—By applying Rule 1a to each of the letters P, Q, and R we can rewrite the expression as shown next. We use a right arrow (→) to indicate that one of the rules has been applied to show that some part of the expression has been proven to be a proposition. Notice that we are not trying to determine what the propositions *means* but merely trying to determine if the proposition is *properly written*.

P and Not (Q or R) → ☐ and not (☐ or ☐)

Step 3—We must now see if any further rules can be applied to what remains of the expression. At this point in our proof, there is only a single rule that can be applied. Rule 2b says that "☐ or ☐" is a proposition and hence we can replace the expression "☐ or ☐" with a box.

☐ and not (☐ or ☐) → ☐ and not (☐)

Step 4—We must continue to find further rules to apply to what remains of the expression. Again, there is only a single rule that can be applied. Rule 2f says that "(☐)" is a proposition and hence we can replace the expression "(☐)" with a box since it is a proposition.

☐ and not (☐) → ☐ and not ☐

Step 5—We now notice that we can replace "not ☐" with a box since "not ☐" is a proposition as defined by Rule 2e.

☐ and not ☐ → ☐ and ☐

Step 6—Finally, it should be obvious that Rule 2a can be applied to complete our proof. We can replace "☐ and ☐" with a box to yield.

☐ and ☐ → ☐

At this point we have collapsed the original expression into a single box. This shows that the original combination of letters, words, and parentheses is, in fact, a well-formed and meaningful proposition. When constructing our proof, every time that we applied one of the rules to replace a phrase with a box we identified a proposition that was a smaller part of the larger whole. Figure 3.1 shows how our proof can be drawn as a series of boxes that are nested within one another.

Some mixtures of letters and operators might appear to be meaningful Boolean propositions but are really just meaningless jumbles of symbols. Consider, for example, the following expression.

P not Q

FIGURE 3.1 The Boolean proposition "P and not (Q or R)" is shown to be well formed such that each dotted box corresponds to some proposition that is an element of the whole.

If we apply the same process of using rules to replace elements of the expression, we will eventually show that this expression is not well formed.

Step 1—We first recognize that Rule 1 can be applied to the letters P and Q.

P not Q → ☐ not ☐

Step 2—We then recognize that Rule 2d can be applied to "not ☐". Replacing "not ☐" with a box yields the following.

☐ not ☐ → ☐☐

Step 3—At this point we conclude that "P not Q" is not a proposition since there is no rule that will allow us to reduce ☐☐ to a single box.

The phrase "P not Q" is actually a sequence of two separate propositions rather than a single proposition. Although each of the two propositions may be meaningful when considered in isolation, the sequence itself does not have a truth value. We can illustrate the nonsensical nature of "P not Q" by expanding the abbreviations into a propositional statement. If, for example, we understand P to abbreviate "I am hungry" and Q to abbreviate "I am cold" we understand the expression "P not Q" to be something like "I am hungry I am not cold." Just as the rules of English grammar require us to insert a semicolon or perhaps a period between these two propositions, Boolean logic expects to see a logical connective such as "and" joining the two propositions. Since there is no connective, the resulting phrase is nonsensical. Joining the two propositions with a logical operator such as "and" would yield the meaningful proposition "I am hungry and I am not cold."

3.2.2 Evaluating Propositions

When we evaluate a proposition, the goal is not to make sure that it is well written but rather to determine the truth value of the proposition. Consider, for example, the following simple propositions.

P = "I am hungry."

Q = "I am cold."

If we are then asked to evaluate the truth value of the compound proposition given next, we must very carefully apply certain rules of logic to determine the answer.

P and Q

The truth value of the compound proposition "P and Q" depends upon the truth values of both P and Q. Perhaps I have just finished eating a very large dinner of sushi, turkey, fruit salad, and organic beets. In this case, we understand that the statement "I am hungry and I am cold" is false because the simple proposition that "I am hungry" is false. Similarly, perhaps I am resting on the beach during a sizzling summer afternoon. In this case, we understand that the statement "I am hungry and I am cold" is false because the simple proposition that "I am cold" is false. In Boolean logic, simple propositions are variables that affect the truth value of any compound proposition that contains them.

Although it is not overly difficult to determine the truth value of a proposition such as "P and Q," more complex propositions require careful thought to evaluate. The following two propositions illustrate that logical expressions can be complex enough to require significant effort to evaluate.

not P or not Q and R

not P or Q and R or (P and Q)

Fortunately, Boolean logic defines evaluation rules that, when carefully followed, will allow us to determine the truth value of any compound proposition. These rules of evaluation are centrally related to understanding the logical operators *and*, *or*, *implies*, and *not*.

An *operator* is something like a machine that accepts inputs, processes those values, and produces a single output value. The arity of an operator

Operator	Technical Name	Arity	Example of Use
and	conjunction	2 (binary)	P and Q
or	disjunction	2 (binary)	P or Q
implies	implication	2 (binary)	P implies Q
≡	equivalence	2 (binary)	P ≡ Q
not	negation	1 (unary)	not P

FIGURE 3.2 The logical operators.

is the number of inputs into the operator. Each of the operators also has a technical name that describes the underlying function of the operator. Figure 3.2 gives the arity and technical name for each of the logical operators. Operators that have an arity of two are known as *binary operators* and operators that have an arity of one are known as *unary operators*. The truth value that is generated by a logical operator is dependent only on the input(s).

3.2.2.1 Conjunction (AND)

For the binary operators there are only two inputs and since each of these two inputs has only two possible values there are only four situations that the operator must handle. Each of the binary operators can be fully described by naming the truth value that the operator generates for each of the four input conditions. Consider, for example, the following definition of logical conjunction.

1. The logical conjunction of false with false is false.

2. The logical conjunction of false with true is false.

3. The logical conjunction of true with false is false.

4. The logical conjunction of true with true is true.

Although this is a perfectly reasonable way to define the operators, the definition is more wordy than necessary. To be concise, the logical operators are usually defined as a *truth table*. Each row of a truth table corresponds to one combination of inputs, every column except for the last gives the input values for the operator, while the final column gives the truth value for that input combination.

Conjunction

P	Q	P and Q
False	False	False
False	True	False
True	False	False
True	True	True

FIGURE 3.3 Truth table for logical conjunction.

Consider, for example, converting the prose definition of logical conjunction into the truth table as shown in Figure 3.3. Since there are four combinations of P and Q, there are four rows in the table. Since there are two variables, the table has three columns. The first two columns name the first and second inputs, whereas the final column gives the truth value generated by the logical conjunction of the two inputs.

Logical conjunction yields a value of true only when both of the inputs are true. This result is intuitive since it closely follows the way we informally use the term *and* when speaking. We intuitively understand statements like "I am hungry and I am cold" to be true only when the statement "I am hungry" is true and "I am cold" is also true. In every other case, we understand that "I am hungry and I am cold" is false.

3.2.2.2 Disjunction (OR)

Logical disjunction, however, is not as intuitive since our informal use of the word *or* is not as closely connected to the way it is formally defined. We often pose two mutually exclusive possibilities and ask which of those two possibilities actually occurred. We might, for example, ask our friend "Did you watch television or did you eat dinner?" We would expect them to tell us which of these two actions they actually took, and not expect them to reply "I did both" or "I did neither." We would be even more surprised to hear them respond with "Yes. I did watch television or eat dinner." Logically, however, the correct answer to the question you posed would be either yes or no. The answer would be yes if your friend performed either or both of the two actions. The answer would be no if your friend neither watched television nor ate dinner but rather went jogging in the park. Logical disjunction is formally defined by the truth table of Figure 3.4.

Disjunction

P	Q	P or Q
False	False	False
False	True	True
True	False	True
True	True	True

FIGURE 3.4   Truth table for logical disjunction.

Note that we define the result to be true if either of the inputs is true. Said another way, logical disjunction yields a false value only when both of the inputs are false. Given this definition, we now understand a statement such as "I am hungry or I am cold" to be false only when I am neither hungry nor cold. If, for example, I am both hungry and cold, the statement "I am hungry or I am cold" is understood to have a value of true.

3.2.2.3 Implication (IMPLIES)

Logical implication is perhaps the most challenging operator to master. This operator captures the idea that if one thing is true, then some other thing must also, by logical necessity, be true. For example, if the proposition "my car battery is dead" has a value of True, this implies that the proposition "my car won't start" must also be true. In Boolean logic we would say that "My car battery is dead implies my car won't start." We might phrase this informally as "whenever my car battery is dead, my car won't start."

Generally, of course, we would use abbreviations to represent the simple propositions in which case we would write "P implies Q" where P is an abbreviation of the proposition "My car battery is dead" and Q abbreviates "My car won't start." The truth table defining logical implication is given in Figure 3.5.

Logical implication is a very important operator since it has properties that precisely express correct human reasoning. Unfortunately, logical implication is perhaps the most poorly understood of the logical operators. We therefore describe a few important properties of logical implication in the following paragraphs.

In the proposition "P implies Q" we refer to P as the *antecedent* and Q as the *consequent*. Consider the following two simple propositions and their abbreviations.

Implication

P	Q	P implies Q
False	False	True
False	True	True
True	False	False
True	True	True

FIGURE 3.5   Truth table for logical implication.

P = "My car battery is dead."

Q = "My car won't start."

Consider the following compound proposition that establishes simple proposition P as the antecedent to consequent Q.

P implies Q

This compound proposition has a value that is either True or False, and the value of the proposition is dependent on the value of the two simple propositions P and Q. We will carefully consider each of the four possible combinations of inputs starting from the bottom of the truth table and working our way to the top of the table.

1. P is True and Q is True. Consider the situation where we observe that "my car battery is dead" and we also observe that "my car won't start." Given these observations it is logically consistent to claim that "P implies Q" since it appears that the car won't start because the car battery is dead. In this scenario, we understand that "True implies True" is logically consistent with the observed facts and therefore has a value of True.

2. P is True and Q is False. We now observe that the car will start (since Q is False) and that the car battery is dead. The proposition that "P implies Q" must therefore be false since this proposition is claiming that the car won't start whenever the car battery is dead. In this scenario we understand that "P implies Q" is logically inconsistent with the observed facts. Therefore, "True implies False" has a value of False since it expresses the provably incorrect proposition that whenever the car battery is dead the car won't start.

3. P is False and Q is True. When the antecedent is false, there is no conclusion that can logically be drawn with respect to the consequent. In this case, the statement that "P implies Q" has not been disproven and therefore has a value of True. For example, the observation that "my car battery is not dead" and that "my car won't start" does not contradict the statement that "whenever my car battery is dead my car won't start." Perhaps the car won't start for some reason unrelated to the car battery. Perhaps the spark plugs are polluted or the gas tank is empty. Since there is no logical contradiction between the observed situation and the statement that "My car battery is dead implies my car won't start" we understand that "False implies True" is true.

4. P is False and Q is False. Again, since the antecedent is False, we understand that "False implies False" is True. In this situation, we observe that "my car battery is not dead" and we also observe that "my car will start." It should be apparent that the implication is True since there is no logical contradiction between what we observe and the proposition that "My car battery is dead implies my car won't start."

We define the *converse* of "P implies Q" to be "Q implies P." The converse of a proposition is essentially the logical opposite. One of the properties of implication is that if an implication is true, the converse may not be true. If, for example, it is true that "my car battery is dead implies that my car won't start" we cannot logically conclude that "my car won't start implies that my car battery is dead." There are many reasons that my car may not start and only one of those reasons is that my car battery is dead. Consider the implication that "Humphrey is a dog implies Humphrey is a mammal." It is obvious that the converse is not necessarily true since "Humphrey is a mammal" does not imply that "Humphrey is a dog." Although Humphrey might be a dog, Humphrey might also instead be a cat, a lion, a whale, or even a person.

3.2.2.4 Equivalence (≡)
Variables P and Q are said to be equivalent if they have the same truth value. The truth table for logical equivalence is given in Figure 3.6.

3.2.2.5 Logical Negation (NOT)
Logical negation is the only unary operator. This operator expresses the obvious thought that if something is not true then it must be False, and

Equivalence

P	Q	P ≡ Q
False	False	True
False	True	False
True	False	False
True	True	True

FIGURE 3.6 Truth table for logical equivalence.

if something is not False then it must be True. The truth table for logical negation is shown in Figure 3.7.

3.2.2.6 Compound Propositions

When we *evaluate* a proposition, the goal is to determine the truth value of the proposition. If the proposition involves only a single operator, we can simply look up one row of a truth table to obtain the answer. Consider evaluating the proposition

P or Q

for the situation when P = False and Q = True. We can evaluate the value of "P or Q" by scanning through the disjunctions truth table, finding the line that corresponds to P = False and Q = True and then reading the truth value under the column heading "P or Q."

To evaluate propositions that involve more than one operator, we can apply the same process, but it must be applied once for every operator in the proposition. Consider, for example, the proposition

P and (Q or R)

Negation

P	not P
False	True
True	False

FIGURE 3.7 Truth table for logical negation.

for the situation where P = True, Q = False, and R = True. We first note those parentheses are used to group the subproposition "(Q or R)." For this reason, we first use the disjunction table to evaluate "(Q or R)" where Q = False and R = True. The truth table for disjunction states that if the left value is False and the right value is True, then "Q or R" has a value of True.

Once we have evaluated "(Q or R)" to True we can replace that subproposition by its value and continue our process of evaluation. We now have the proposition

P and True

There is now a single logical operator and we complete our evaluation by using the truth table for conjunction to find the value corresponding to the case where the left value is true and the right value is true. We note that the value is true. In other words, the truth value of "True and (False or True)" is true.

When evaluating a proposition that involves more than one logical operator you should normally proceed from left to right. The only exception to this general rule is when parentheses are used to group elements in some other natural order. Consider, for example, the proposition

P and Q implies R

for the situation where P = True, Q = True and R = False. There are two operators in this proposition and they should be evaluated from left to right. The first step is to evaluate the subproposition "P and Q" for P = True and Q = True. We find that "True and True" has a value of True. We now evaluate the expression "True implies R" for R = True. We find that "True implies True" has a value of True and hence the entire proposition "P and Q implies R" has a value of True.

We can use parenthesis to group the operators differently and this affects the order in which the operators should be evaluated. Consider, for example, the different grouping of operators shown next for the situation where P = True, Q = True, and R = False.

P and (Q implies R)

Since the implication operator is in parentheses, we must evaluate that operator first. In our case we discover that "True implies False" has a value of False. We then evaluate the conjunction operator in the proposition "P implies False," or, since P = True, we evaluate "True implies False," which yields a value of False. From this example we see that the meaning of a proposition can be changed by different groupings of the operators.

Truth tables can be used to express the meaning of any logical proposition. To construct a truth table for some proposition you must do the following.

1. Make a list of each logical variable (abbreviation) that appears in the proposition. If one variable occurs more than once in the proposition it should be included only once in your list.

2. Place a column in the truth table for every variable list your list. The column heading should be the variable itself.

3. For the heading of the last column you should write the entire proposition.

4. If there are N variables in your list, you must create 2^N rows. Each row represents a unique combination of values for the N variables.

5. For each row, determine the value of the proposition and place that value in the last column.

Consider, for example, the proposition

P and not (Q or R)

We construct the truth table by first making a list of the logical variables listed in the proposition. The logical variables P, Q, and R are the only variables that occur and so our list contains only those three variables. We now construct a truth table that has one column for each of these variables. In addition, the final column corresponds to the whole proposition. Since there are three variables, we now construct 2^3 or 8 rows such that each row corresponds to a unique combination of values over the variables P, Q, and R. This table is shown in Figure 3.8.

It is preferable to arrange the rows in some orderly fashion. In this example, you should note that the rightmost column alternates repeatedly between True and False. The next column to the left alternates in a pattern of two False values followed by two True values. The next column to the left alternates at a rate that is half as frequent as the previous column.

P	Q	R	P and not (Q or R)
False	False	False	
False	False	True	
False	True	False	
False	True	True	
True	False	False	
True	False	True	
True	True	False	
True	True	True	

FIGURE 3.8   Truth table for a logical proposition.

The challenge is to complete the table such that the value of the proposition "P and not (Q or R)" is known for each possible combination of P, Q, and R. In order to determine these values it is convenient to temporarily include columns for each of the logical operators that must be evaluated to determine the value of the whole proposition. These subparts should be arranged according the order in which the operator will be evaluated. Figure 3.9 gives an example of such a table.

P	Q	R	(Q or R)	not (Q or R)	P and not (Q or R)
False	False	False	False	True	False
False	False	True	True	False	False
False	True	False	True	False	False
False	True	True	True	False	False
True	False	False	False	True	True
True	False	True	True	False	False
True	True	False	True	False	False
True	True	True	True	False	False

FIGURE 3.9   The process of generating the truth table.

P	Q	R	P and not (Q or R)
False	False	False	False
False	False	True	False
False	True	False	False
False	True	True	False
True	False	False	True
True	False	True	False
True	True	False	False
True	True	True	False

FIGURE 3.10   Final truth table for the proposition "P and not (Q or R)."

The proposition contains three logical operators and hence there are three columns to the right of the variables. Two of these columns—the columns corresponding to disjunction and negation—are temporary and will not be retained as part of the final table. The final column is generated by the conjunction operator and this column will be retained since the conjunction operator is the last operator evaluated and generates the value of the proposition. Eliminating the temporary columns yields the truth table of Figure 3.10.

3.2.2.7 Logical Equivalence

Two propositions are said to be *equivalent* if they have the same truth table. Consider, for example, the two propositions

not (P and Q)

(not P) or (not Q)

The truth table for each of these propositions is listed in Figure 3.11.

Since these two tables are identical, we know that the proposition "not (P and Q)" is equivalent to the proposition "(not P) or (not Q)." When two propositions are equivalent, they have the same meaning and are therefore interchangeable.

3.2.2.8 Tautologies and Contradictions

Any proposition that has a value of True regardless of the inputs is said to be a *tautology*. Any proposition that has a value of False for all inputs is said

P	Q	not (P and Q)	P	Q	(not P) or (not Q)
False	False	True	False	False	True
False	True	True	False	True	True
True	False	True	True	False	True
True	True	False	True	True	False

FIGURE 3.11 Two different propositions have the same truth table. These two propositions are therefore said to be equivalent.

to be a *contradiction*. Since a tautology is always True, a tautology can be understood as simply a long-winded way of expressing something that is self-evident. Of course, a contradiction can be understood as a statement that is always logically invalid no matter how you look at it. Tautologies and contradictions should be avoided when writing propositions and they should also be avoided in the normal course of human conversations since such statements do not express any sort of productive line of reasoning. The following propositions give examples of a tautology and a contradiction.

Tautology: P or not P

Contradiction: P and not P

Tautologies are often used in political debate and political rhetoric. Consider, for example, the statement, "If you elect me to be president then I will serve as the elected president." This statement is simply a long-winded statement of the obvious. When expressed as a logical proposition, the statement reduces to "P implies P," which can be easily shown to be a tautology by constructing the truth table.

When we think that our friend (or even our teacher) is saying something that does not make sense it is often because we think that they are making a contradictory statement. Consider the following conversation between two students in a computational thinking course.

Matthew: My new Internet connection is blazingly fast. I watched HD movies all night.
Susan: Have you finished your computational thinking homework?
Matthew: No. My Internet connection is so painfully slow that I can't download the one-page assignment.

We intuitively understand that Matthew is not making sense because he has expressed a contradiction. The source of the contradiction is his statement that his Internet connection is both "blazingly fast" and "painfully slow" at one and the same time. We can express this contradiction as the contradiction "P and not P," where P abbreviates the statement "My Internet connection is fast." If a line of human reasoning is ever shown to exhibit a contradiction, it is certain that the line of reasoning is logically flawed. Valid reasoning requires that a proposition cannot be both true and false at the same time; this requirement is referred to as the *law of noncontradiction*.

3.3 APPLICATIONS OF PROPOSITIONAL LOGIC

Although you might expect physicists, mathematicians, and accountants to use formal logic in their daily profession, you may be surprised to discover that a much wider range of disciplines also make common use of propositional logic. In this section we will describe how propositional logic is used by anyone who searches the Internet, electrical engineers, database designers, and even graphic artists!

3.3.1 Search Queries

A web search engine is a software system to search for information on the World Wide Web. Search engines work by downloading web pages and recording each word or phrase that occurs on the page. Whenever users want to know where they can find information about a topic, say "logic," they can type in the word "logic" and the search engine will return a list of all web pages that contain the word "logic."

A *search query* is a phrase that describes the information that the user seeks to obtain. Many search queries are a single word such as "rose," "logic," or "platypus." Although single-word queries are sometimes effective, they often fail to produce the right type of information because the word may be ambiguous or frequently used across many different contexts.

Consider a situation where you live in a city that has a grocery store named "People." You would like to use a search engine to find the phone number and the hours of operation for this store. You naturally use the term "people" as a search query but find that your query does not produce useful results. When this text was written, the top search engines returned results such as the *People Magazine* website, Yahoo's People search page, the website of "Peoples Bank," a Wikipedia article titled

"People," a Wikipedia article devoted to a band named "Peoples," and a link to a web page devoted to the "People of Walmart."

Simple single-word queries are often not specific enough to provide useful results. Most search engines therefore allow you to construct more specific search queries that narrow the focus of your query in a more useful manner. The Boolean operators AND, OR, and NOT are one technique that is commonly used to narrow a search query.

3.3.1.1 Conjunction in Search Queries

By default, all search queries that use more than one word use conjunction. In other words, the search query returns web pages that contain all the terms or phrases in the query. As an example, consider searching for information about a high school classmate named Hannah Garcia. You first decide to use the search query "Hannah" but the query returns a flood of web pages related to Hannah's of all kinds; none of whom are related to the "Hannah Garcia" that was your high school friend. You decide to narrow your search by forming the two-word query "Hannah Garcia." The search engine understands this query to be the conjunction of the two words. In other words, the search engine understands the query to mean "Hannah AND Garcia." The only pages that are returned by the search engine are those web pages that contain both the word "Hannah" and the word "Garcia."

3.3.1.2 Disjunction in Search Queries

Unfortunately, you still cannot find any web page related to your high school classmate Hannah Garcia. As you think about how to improve your search results, you remember that she married someone with the surname Mason. You realize that although she may still use the surname Garcia in certain business contexts, she may instead use the surname Mason. You therefore attempt to construct a search query that expresses the possibility that her last name is either Garcia or Mason. You initially think that the search query "Hannah Garcia Mason" would be appropriate, but this search query would be overly narrow since only pages containing all three words would be returned as the search engine would interpret this query to mean "find all web pages that contain the words Hannah AND Garcia AND Mason."

After some study you discover that search engines use an OR operator to express alternatives. The alternatives in this case are "Garcia" and "Mason." You therefore construct the search query "Hannah AND

(Garcia OR Mason)." This query correctly expresses your desire to find web pages that contain only one of the surnames Garcia or Mason but not both.

3.3.1.3 Negation in Search Queries

Perhaps the search query "Hannah AND (Garcia OR Mason)" returns many web pages related to a well-known soccer player. You are certain that your high school classmate is not a soccer player and therefore you attempt to construct a search query that prevents soccer-related pages from appearing.

Search engines use the NOT operator to exclude pages that contain certain words. You therefore construct the following search query: "Hannah AND (Garcia OR Mason) AND NOT soccer". This query will find all pages that contain either the words "Hannah" and "Garcia" or the words "Hannah" and "Mason," while preventing any web page containing the word "soccer."

Negation is expressed in some search engines as a dash (–) immediately preceding the word that should be excluded. In this case, the search query "Hannah AND (Garcia OR Mason) AND NOT soccer" would be expressed as "Hannah AND (Garcia OR Mason) AND –soccer". Also note that most search engines require the logical operators to be entered in all capital letters. If the word is typed in lowercase, the search engine will likely return all web pages related to the word "and," for example, rather than interpreting AND as an operator!

3.3.2 Digital Logic

Digital circuits are the physical components of a computing system. A digital circuit is an electronic system that enables a computer to perform arithmetic operations such as addition and multiplication among many other operations. These digital circuits are typically constructed by combining logic gates in various ways. A *logic gate* is an electronic device that implements a Boolean operator. In other words, a logic gate will have inputs and produce a single output corresponding to the operator that it implements.

Figure 3.12 shows two electrical circuits that implement the Boolean operators AND and OR. For each case, a battery is connected to a lightbulb and two switches. For the AND circuit, it is apparent that both switches must be closed in order to light up the bulb. If either one of the switches in the OR circuit is closed, the lightbulb will be illuminated.

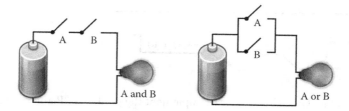

FIGURE 3.12 Electrical circuits where switches describe the logical input and the state of the lightbulb denotes the output.

Logic gates are usually implemented using transistors as switches. In the context of digital logic, a battery can be thought of as a power source with a voltage level of 1. Since the switches either supply full power or no power to the output device, there are only two signal levels present in the circuit. These two signals are usually denoted using 0, which corresponds to no power or False and 1, which corresponds to full power or True. These logic gates are usually drawn using the symbolic forms of Figure 3.13.

While a simple proposition can be represented by one of these logic gates, compound propositions can be represented as a combination of these simple logic gates. Digital circuits are therefore equivalent to compound Boolean propositions. Consider using logic gates to implement the compound predicate

P and not (Q or R)

Since the "or" operator is enclosed in parenthesis, that part of the expression is performed first. In a logic circuit, this implies that the "or" gate occurs first in the path of the signal. The resulting expression, (Q or R) is then inverted and finally fed into the "and" gate to produce the result shown in Figure 3.14.

When designing an electronic circuit we would prefer to use as few gates as possible. Since each gate increases the cost of the system, minimizing the number of logic gates results in a less expensive system. Also, fewer gates often also means that the circuitry is faster since a digital signal

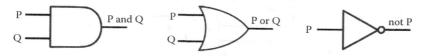

FIGURE 3.13 Schematic notation for the Boolean logic gates AND, OR, and NOT.

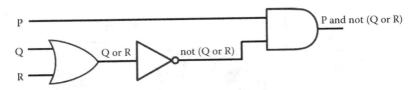

FIGURE 3.14   Using logic gates to implement the predicate "P and not (Q or R)."

does not need to travel through as many gates. Since digital circuits are expressions of Boolean propositions, it is natural to use Boolean algebra to design and analyze complex digital circuitry.

One final example of a basic digital circuit is a circuit that performs an equivalency check on two inputs. Figure 3.15 has an output of True whenever P = Q, otherwise the output is False. This circuit encodes the concept that P is only equal to Q whenever (a) both P and Q are True or whenever (b) both P and Q are False.

3.3.3 Image Compositing

A graphic designer is a professional who works with images and fonts to create the graphics for such things as textbooks, brochures, movies, and web content. Since the core responsibility of the designer's job is to present visual information in a compelling manner, graphic designers are highly skilled at creating and editing digital images. Many important editing functions used daily by the graphic designer are direct applications of formal logic.

Digital images can be thought of as a grid of small, colored cells known as pixels that when viewed from a sufficient distance will blend together to form a high-quality picture. *Binary images* are digital images composed of only white or black pixels. Although binary images are not colorful, they nonetheless can convey a great deal of information and visual detail.

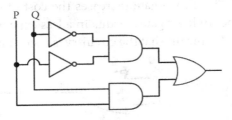

FIGURE 3.15   Equivalence circuit. The output is 1 (True) whenever P is equivalent to Q.

FIGURE 3.16   Binary image of the *Mona Lisa*.

For example, consider the image of Figure 3.16 that shows a binary image derived from the *Mona Lisa*.

Since there are only two colors in a binary image and only two logical values, we can associate a color with a logical value. For this discussion we will associate the color black with logical value True and the color white with the logical value False. We can now apply logical operations on colors using the logical operators AND, OR, IMPLIES, EQUALS and NOT. Since the logical values are associated with colors, the truth tables for these operators can be expressed using the colors black and white as shown in Figure 3.17.

Graphics software systems give artists the ability to construct complex shapes and pictures by combining simpler shapes. *Digital compositing* is the process of combining two shapes or images to create a more

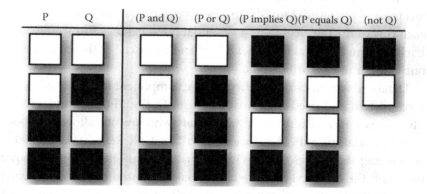

FIGURE 3.17   Truth tables using colors.

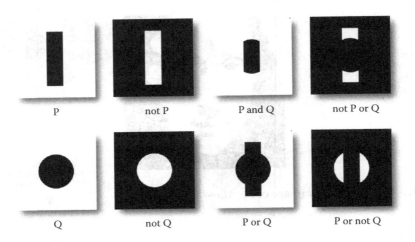

FIGURE 3.18   Image compositions using logical operators.

complex image. Some compositing techniques are direct applications of propositional logic where colors are used in place of truth values.

If we assume that P and Q are binary images, we can then create composite images by writing a proposition that includes logical variables P and Q. These composite images are constructed by applying some logical operator to every pixel in the images that are being composited. Figure 3.18 gives an example of this technique. Images P and Q can be composited and expressed as predicates to form a variety of other binary images.

3.3.4 Database Queries

Databases are software systems designed to efficiently store enormous amounts of data such that pieces of data in the database can be very quickly located and retrieved. Most databases store information in tables such that each row of the table contains a set of data that belongs to a single *record*. Each column of the table defines a *field* and every cell of the table contains one field for one record of data.

Database systems are an indispensable component of almost every software system and used in almost every profession. For example, the No Fly List is a large database of individuals who are not allowed to board a commercial aircraft, the National Do Not Call Registry is a database containing the names of individuals that telemarketers are not allowed to solicit, the National Sex Offender Registry is used by law enforcement to track the location of convicted sex offenders, every bank and financial institution maintains a database of clients and accounts, Wikipedia maintains a database of articles, biologists maintain large databases of DNA sequences, and politicians maintain large databases of registered voters.

Consider the case of a political lobbyist who wants to raise funds for some political cause. The lobbyist wants to concentrate all of their fund-raising efforts on a narrow group of likely donors and has been given access to a database of individuals who have made similar donations in the past.

The structure of the database is given in Figure 3.19. Since there are six columns we understand that each record consists of six fields. Those fields are named First, Last, Age, Amount, State, and Party. There are seven rows and hence we understand that the table contains seven records. In this case, each record contains a person's name, the amount donated, their state of residence, and their party affiliation. The political party is denoted as D for Democrats, R for Republicans, T for the Tea Party, and I for all independents.

Perhaps the lobbyist wants to solicit individuals who have made similar donations of at least $500. The list of potential donors can be generated by selecting a subset of the individual donors in the database. In this case, the lobbyist will generate this list by scanning down the column labeled Amount and selecting those rows having a value of at least 500 in the amount column. Since there are only seven records in our naïve example, generating the list of potential donors is easily done, but most real-world databases contain many millions of records. The enormous size of data in

Donor Table					
First	Last	Age	Amount	State	Party
William	Shell	61	1,300	WI	D
Lisa	Dough	19	125	WY	T
Helen	Lobby	35	1,200	CA	R
Reggie	Green	33	800	NY	R
Harvey	Levirage	41	500	NY	I
Robin	Round	32	200	NY	D
Jennifer	Dichali	38	650	WI	D

FIGURE 3.19 A relational database table.

most real-world databases requires sophisticated software to extract and analyze data of interest.

Fortunately, databases are designed to efficiently locate records in response to well-formed queries. Although somewhat similar to a web search, a database query requires more precise structure and is generally written in a language known as *SQL*, which is short for *Structured Query Language*. Records in a database are located by issuing a select statement. The general form of a select statement is given as

SELECT *field1, field2, …, fieldN* FROM *table* WHERE *criteria*

where the italicized elements represents the details that control what data is retrieved. Consider writing an SQL query to select the names of donors who have contributed at least $500. Since we are only interested in the names of the individuals, we would select only the First and Last fields. We are selecting data from the Donor table and the criteria that determines which elements to select is expressed as Amount >= 500. The proper SQL query is given as

SELECT First, Last FROM Donor WHERE Amount >= 500

The criterion of this select statement is a simple logical predicate that locates and retrieves only those records of interest to the lobbyist. The result of the select statement is a list of five donors: William Shell, Helen Lobby, Reggie Green, Harvey Levirage, and Jennifer Dichali. Consider, however, a scenario where the lobbyist wants to specifically target only those donors who are under 40 years of age and have contributed at least $500 in the past. This requires a more sophisticated criterion; a criterion that is expressed as a compound logical predicate and produces a list of the three donors: Helen Lobby, Reggie Green, and Jennifer Dichali.

SELECT First, Last FROM Donor WHERE Amount >= 500 AND Age
    < 40

Perhaps the lobbyist wants to specifically target only those donors who live in either the coastal states of New York or California regardless of their age or their past donations. The appropriate SQL query is expressed as

SELECT First, Last FROM Donor WHERE State = 'NY' OR State = 'CA'

Note that the terms NY and CA must be enclosed in quotations to distinguish those elements as merely textual elements. This query produces a list of four donors: Helen Lobby, Reggie Green, Harvey Levirage, and Robin Round.

For our final example, consider an SQL query to locate every donor that is under 40 years of age, has donated at least $500 in the past, and lives in either New York or California. The table contains only two such donors: Helen Lobby and Reggie Green. The SQL query to generate this list is given as

SELECT First, Last FROM Donor WHERE Age < 40 AND Amount >= 500 AND (State = 'NY' OR State = 'CA')

The ability to precisely express database queries as logical predicates is a vital skill for anyone who works with large amounts of structured data.

3.3.5 Software Requirements

When a software engineer is hired by a client to write software, the first and often the most difficult aspect of the job is to define as precisely as possible what the software should do. The process of defining what a piece of software should do is known as *requirements engineering*. Typically, the software engineer will schedule a series of meetings with the client and spend hours talking to everyone interested in the software that is being created. The software engineer will make sure that every detail of everything that is discussed in these meetings is written down and precisely defined so that there is one common understanding of every aspect of the system.

The end result of requirements engineering is a document known as the *software requirements document*. This document is a very precise definition of what the software should do once it has actually been written. The software requirements document often serves as a formal contract between the client and the software engineer, and hence both the client and the software engineer must both agree that the software requirements document is a precise definition of what the software should do. This necessitates that the document should also written in such a way that both the client (often a nontechnical person or company) and the engineer are able to fully understand.

Logical propositions are often used to define important functions of a large software project. Consider, for example, a corporate client that has hired a software engineer to write software to control an elevator system for a small four-floor office building. The elevator has two doors that are known as the front and the rear door. The front doors are usable only on the first two floors, and the rear doors are usable only on the top two floors. An interview session between Germaine, the software engineer, and Aditi, the main corporate contact, might go something like the following dialogue.

Germaine: What should happen when the "open door" button is pressed?
Aditi: The door should open.
Germaine: The front door or the rear door or both?
Aditi: The front door should open if the elevator is on either floor 1 or 2, otherwise the rear door should open.
Germaine: What if the "open door" and "close door" buttons are pressed at the same time?
Aditi: Hmmm. I hadn't thought about that. I guess nothing should happen.
Germaine: What if the elevator is moving between two floors and the "open door" button is pressed?
Aditi: Nothing should happen.
Germaine: What if emergency personnel have inserted a key to manually override the software control system. In that case, what should happen when the "open door" button is pressed?
Aditi: Hmmmm. That's a good question. I guess that both the front and rear doors should open regardless of which floor the elevator is on or even if the elevator is moving between floors.

After several discussions, Germaine will then include a specification for the "open door" function of the elevator. This function will likely involve a number of logical predicates that very precisely define what should happen when the "open door" button is pressed. The functionality might be described in the following manner.

Description: The elevator contains both an "open door" button and a key slot that is intended for the use of emergency personnel. The following specification defines what the control software should do when the "open door" button is pushed.

- (FLOOR=1 OR FLOOR=2) AND (NOT MOVING) AND BUTTON_PUSHED IMPLIES FRONT_DOOR_OPENS

- (FLOOR=3 OR FLOOR=4) AND (NOT MOVING) AND BUTTON_PUSHED IMPLIES REAR_DOOR_OPENS
- (EMERGENCY_KEY_INSERTED AND BUTTON_ PUSHED) IMPLIES (FRONT_DOOR_OPENS AND REAR_DOOR_OPENS)
- In all other cases, the system must do nothing

Software systems that might cause significant harm if they operate improperly are known as *safety-critical systems* and are specified using very precise mathematics. Software that controls an artificial heart, an insulin pump, a medical CT scanner, a national air-traffic control system, the automatic pilot software on a large commercial jetliner, or the control software of a nuclear power plant are all known as safety-critical systems. *Formal methods* are a set of mathematically rigorous techniques for the specification and development of safety-critical software systems. Although most software systems are specified using a mixture of mathematical formalisms and nonmathematical descriptions, the more critical the system, the more formal the specification must be.

TERMINOLOGY

and	database
antecedent	deductive logic
binary image	digital compositing
binary operator	digital image
Boolean logic	disjunction
compound proposition	equivalence
conclusion	evaluate
conjunction	field
consequent	implication
contradiction	implies
converse	inductive logic

law of noncontradiction

law of the excluded middle

logic

logic gate

logical values

negation

not

operator

or

premise

proposition

record

requirements engineering

search query

simple proposition

software requirements document

SQL

structured query language

syllogism

symbolic logic

tautology

truth table

truth values

unary operators

well-formed propositions

EXERCISES

1. Name the two logical values.

2. For each item, identify whether the item is a well-formed proposition.

    a. P NOT Q

    b. NOT P OR Q

    c. NOT P NOT Q

    d. P AND NOT Q

    e. P AND OR Q

    f. NOT P AND NOT Q

3. Give a well-formed proposition that is equivalent to (P IMPLIES Q).

4. Give the truth table for each of the following propositions.

    a. P AND (Q IMPLIES R)

    b. (P OR Q) AND R

    c. P OR (Q AND R)

    d. NOT (P AND (Q OR R))

5. Give an interesting web query that uses logical conjunction, disjunction, and negation in a substantive manner to produce an interesting result.

6. Draw a digital circuit for each of the following propositions.

    a. (P OR Q) AND R

    b. P OR (Q AND R)

    c. NOT (P AND (Q OR R))

7. Draw a digital circuit for the proposition that (P IMPLIES Q)

8. Consider the following database table. The database is used by a local falconry club to track its members. The Paid column indicates whether a person has paid dues for this year, the Gender column has the entry U for those members who have signed up but not yet provided their gender information. For each of the following criteria, write an SQL query, following the examples given in this chapter, to generate the desired result.

    a. The first and last name of all female members

    b. The e-mail addresses of those members who have paid their dues

    c. The e-mail addresses of those members under 21 who have not paid their dues

    d. The first and last name of all members who are either male or have an unknown gender

    e. The e-mail address of all dues-paying members who have an unknown gender

Member Table					
First	**Last**	**e-Mail**	**Paid**	**Gender**	**Age**
Jason	Lin	jlin@ctmail.com	Y	M	32
Hannah	Flynn	fly@ctmail.edu	N	F	19
Jordyn	Ash	ash@ct.com	Y	U	59
Reginald	Holt	reholt@ct.edu	Y	M	31
Sophia	Grace	sophi@ct.edu	Y	F	29
Ella	Peters	ellap@ct2 .edu	N	F	23
Dakota	Flynn	dfly@ctmail.edu	N	U	19
Aaron	McGregor	amg@ctmail.edu	Y	M	59

CHAPTER 4

Solving Problems

It's not that I am so smart, it's just that I stay with problems longer.

—ALBERT EINSTEIN

OBJECTIVES

- To examine problem definition as the beginning of all good problem solving
- To explore functional requirements as the core of algorithmic problem definitions
- To examine several ways that logical reasoning is applied to software development, including cause-effect relationships, deductive reasoning, and inductive reasoning
- To understand that programming is an activity that often relies upon knowledge of patterns and that the five basic patterns of control flow are sequences, selection, repetition, control abstraction, and concurrency
- To understand the central role algorithms play in computational problem solving and explore many forms of algorithms
- To examine divide and conquer as a key problem-solving strategy, useful in outlining and top-down design
- To explore prototyping as a form of decomposition that is well suited to solving many of today's problems
- To consider the impact of alternative approaches to data decomposition, such as linear searching as opposed to binary searching
- To explore many forms of abstraction that are significant to computer science, including control abstraction, class diagrams for data abstraction, and use-case diagrams for definition abstraction

It has been said that computer scientists are modern-day problem solvers. So it is impossible to understand computational thinking without understanding the problem-solving skills and techniques of the computer scientist. Of course, problem solving did not begin with computers, nor are computers essential to solve many problems; and, of course, the problem-solving skills common to computer scientists are useful far outside the world of computers.

It would be impossible to capture all the ways that computer scientists solve problems, but there are four strategies regularly employed by computer scientists that are at the core of this modern style of problem solving:

1. *Problem definition*

2. *Logical reasoning*

3. *Decomposition*

4. *Abstraction*

In addition to their significant utility in computational problem solving, each of these techniques has numerous applications aside from computer science.

4.1 PROBLEM DEFINITION

Like the scientific investigation process begins with a hypothesis, software development begins with a careful *problem definition*. The problem definition specifies what task(s) are to be performed by the associated software. The problem definition also serves as the software developer's goal. Without such a goal it is impossible to know whether the problem has been solved, impossible to say whether a computer application is correct.

Software engineers recognize at least three major phases to the process of developing software:

1. Analysis

2. Design

3. Implementation

The latter two phases involve the creation of a problem solution, but *analysis* is all about defining the problem. In some sense analysis is the most important phase, because, as every good software engineer knows, successful design and implementation is only possible given adequate analysis.

Problem definition involves more than just computer scientists. Anyone who contracts the creation of software (that person is known as the *customer* in software engineering terminology) must assist in analysis in order to convey the intended problem definition. In fact a problem definition is the core of the legal agreement between customer and software developer.

A properly engineered software problem definition consists primarily of a list of *requirements*. As the name implies, a requirement specifies one essential aspect of the software. Requirements are further divided into two types: (1) *functional requirements* to specify the particular task(s) the software must perform and (2) *nonfunctional requirements* to define other characteristics and constraints related to the software. Nonfunctional requirements include expectations for things like reliability, safety, security, performance, delivery, and help facilities. These nonfunctional requirements are useful; however, this presentation focuses on functional requirements because these form what most of us think of as the software problem definition.

As an example of functional requirements, consider a computer application to play videos. We will call this application *Video Player*. Informally, this application plays a video with buttons that allow the user to play or pause the video, as well as to raise or lower the sound volume. Figure 4.1 contains some of the functional requirements for this video player. The sample functional requirements are numbered V1 through V4. In this case each functional requirement corresponds to the action of a single user button shown to the right of the requirement.

A good list of functional requirements must be

- Clear

- Consistent

- Complete

Index: *V1*

Name: *Play*

Action: Clicking the play button (see image to the right) causes the video to begin playing. If the video was just paused, then playing the video resumes at the point of pausing. If the video has yet to start playing, then a new computer window is created and the video begins playing in the window. So long as the video is playing the pause/play button functions as a pause button, displaying a pause button image.

Index: *V2*

Name: *Pause*

Action: Clicking the pause button (see image to the right) causes the video to pause at the current play location. If the video has already finished playing, then this button has no effect. Clicking the pause button causes the pause/play button to function as a play button, displaying a play button image.

Index: *V3*

Name: *Increase Volume*

Action: Clicking the increase volume button while the volume is less than the maximum 80 dB level causes the volume to increase by 5 dB Clicking the increase volume button while the volume is at the 80 dB level does nothing.

Index: *V4*

Name: *Decrease Volume*

Action: Clicking the decrease volume button while the volume is greater than the silent (0 dB) level causes the volume to decrease by 5 dB Clicking the decrease volume button while the volume is at the silent level does nothing.

FIGURE 4.1 A few functional requirements for a video player app.

A clear requirement is one that is easily understood in the same way by all customers, users, and developers. Clear requirements need to be unambiguous and precise. Generally, a good functional requirement can be translated into a logical proposition. For example, in requirement V2 the statement, "If the video has already finished playing, then this button has no effect" can be translated into the expression *F implies N*, where *F* denotes the condition that the video has finished playing and *N* denotes a no change kind of action.

Related to the property of clarity is the need for functional requirements that are consistent. Consistent requirements are those that do not contradict one another. In the example, it would be inconsistent to include a fifth requirement that states, "When the video is playing and reaches the end, it immediately restarts playing with the play/pause button functioning as a play button." This statement is inconsistent with V1, because it contradicts V1's specification that the play/pause button should function as a pause button when videos are playing.

The third property of good functional requirements—completeness—is perhaps the most elusive. It takes careful consideration to be certain that every possible scenario has been considered and explained in one or more requirement. Two scenarios that should always be included for completeness are how the application begins and ends its execution. The requirements from Figure 4.1 could be considered incomplete for overlooking these situations. Figure 4.2 provides two more functional requirements to rectify this deficiency.

Another way that functional requirements may lack completeness is when they fail to consider all possible combinations of situations. A good way to analyze for adequate completeness is to construct a *state-activity table* that lists all possible application states against all possible user actions. Figure 4.3 illustrates with such a table for the Video Player.

Index: *V5*

Name: *Run Application*

Action: The Video Player application starts executing automatically whenever the user double-clicks on an associated video file. As this execution begins, the application displays a window containing the control panel, like the image to the right. This control panel remains until the user chooses to quit the application. If the application is executing, other video file double-clicks are ignored.

Index: *V6*

Name: *Quit Application*

Action: Anytime that the Video Player application is executing the user may double-click within the region of the video display. Such a double-click causes the application to create a pop-up window with two buttons—*QUIT* and *CONTINUE*. If the user clicks the *QUIT* button, then the application quits executing. If the user clicks the *CONTINUE* button, then the video continues

FIGURE 4.2   Functional requirements to start and quit the video player app.

User Actions / Application States	App Not Executing	Video Paused	Video Playing	Video Finished
Click Play		V1		
Click Pause			V2	V2
Click Increase		V3	V3	V3
Click Decrease		V4	V4	V4
Double-click video file	V5	V5	V5	V5
Double-click video window		V6	V6	V6

FIGURE 4.3   A state-activity table for the video player app.

The rows of a state-activity table come from potential user actions. The user can click any of the four buttons (i.e., play, pause, increase volume, or decrease volume). The other two user actions are to double-click a video file icon or to double-click an active video display window.

The columns of the table represent the different potential states of the application. The table in Figure 4.3 identifies four different states: (1) the Video Player application has not yet started to execute; (2) the application is executing but currently paused; (3) the video is playing; or (4) the video has been playing but reached the end of the video.

The state-activity table sometimes contains cells that are impossible. In the table of Figure 4.3 such impossible cells are indicated by a gray color. For example, the Video Player control panel is not visible until the application begins to execute. Therefore, it is impossible for the user to click any of the buttons of the control panel while in the App Not Executing state. The remaining impossible states occur because of the way the control panel works. Functional Requirement V1 specifies that when a video is playing the pause function is active, making it impossible to click the play button. Similarly, Requirement V2 states that clicking the pause button causes the control panel to display on the play button, making another click of the pause impossible.

Once the impossible cells of the state-activity table have been determined, the remaining cells should all be fully defined by one or more functional requirement. Figure 4.3 illustrates by entering the number for each requirement within the cell(s) to which it applies. In other words Requirement V6 defines what occurs when the user double-clicks the video display window when the video is paused, playing, or finished.

A state-activity table for a complete set of functional requirements contains requirements in every cell except those that are impossible. Figure 4.3 shows that there is something incomplete about the requirements for the Video Player application, because the upper right cell is blank. The problem is that the requirements do not really explain what occurs when the video has played to completion and the user clicks the play button. (This is possible, because the user could have clicked the pause button after the video finished, and Requirement V2 specifies that this would cause the control panel to display the play button.) Perhaps the play button should restart the video; perhaps it should terminate the application; or perhaps the play button should do nothing in the video finished state. The issue is that the Video Player functional requirements need to be altered to explain the desired functionality in order for these requirements to be considered complete.

4.2 LOGICAL REASONING

The second of the four important problem-solving strategies for a computer scientist is *logical reasoning*. In Chapter 2 we discovered the close relationship of logic and computing in part resulting from representing true or false as a single bit of computer memory. But the usefulness of logical reasoning extends well beyond the obvious connection to computer hardware. In fact, logic plays a role throughout the problem-solving process of software development.

As explained in Section 4.1, the functional requirements of a good problem definition are English forms of logical statements. This is evident in the use of English constructs such as *if … then* that translates into the logical *implies* operation.

Logic, as used in functional requirements and essential at the start of software problem solving, is equally necessary at the end. Virtually all computer software depends upon some form of logical instructions because an executing program must make choices. The program controlling an online website must choose when to bill a customer. The program in your digital camera chooses how long to leave the shutter open when taking a picture. The program controlling your portable music player chooses which song to play next. Each of these choices is accomplished with computer instructions that are a programmatic form of logic. Most programming languages use a so-called *IF instruction* for this purpose:

- IF the user has struck the Confirm Purchase button, THEN initiate billing procedures.

- IF the camera sensor detects adequate exposure, THEN close the camera shutter.

- IF a song has just finished playing, THEN begin playing the next song in the playlist.

IF instructions force a programmer to think in logical ways, to consider a computer program as a complex collection of *cause-effect relationships*. A cause-effect relationship consists of a logical condition (the *cause*) that forces the program to perform some task (the *effect*). Figure 4.4 contains four cause-effect relationships commonly used by computer programs. Much of the work of computer programming is identifying these kinds of cause-effect relationships and translating them into instructions.

The ability to apply general rules to particular situations is known as *deductive reasoning*. This kind logical reasoning was made famous in the

Cause	Effect
User name and password have been entered	Check to see if user's name and password are valid
User clicks the close button on a window	Remove the closed window from the display
Laptop battery charge below 10%	Pop up a window warning user of low power
A new e-mail message arrives from the Internet	Ring a bell to notify the user

FIGURE 4.4 Some typical cause-effect relations from programs.

fiction of Sir Arthur Conan Doyle regarding a master at deduction named Sherlock Holmes. Following is an excerpt from *The Sign of Four* [1]. In this quote, Sherlock Holmes is talking to Dr. Watson explaining how he deduced that Watson had been to the post office recently.

> "It is simplicity itself," he [Holmes] remarked, chuckling at my surprise,—"so absurdly simple that an explanation is superfluous; and yet it may serve to define the limits of observation and of deduction. Observation tells me that you have a little reddish mold adhering to your instep. Just opposite the Seymour Street Office they have taken up the pavement and thrown up some earth which lies in such a way that it is difficult to avoid treading in it in entering. The earth is of this peculiar reddish tint which is found, as far as I know, nowhere else in the neighborhood. So much is observation. The rest is deduction."

In this example, Holmes has deduced that Watson went to the post office by using rules such as (1) walking on freshly thrown earth tends to stick to shoes and (2) eliminating all other possibilities makes the remaining one true. Holmes applied these rules to this specific case, as do modern-day forensic experts who apply laws of science.

Deduction is often used when applying a mathematical axiom, law, or theorem to a particular case. For example, when considering a right triangle the Pythagorean theorem states that $A^2 + B^2 = C^2$, where A and B are the lengths of the triangle legs and C is the length of the hypotenuse. From the Pythagorean theorem we can deduce the distance from opposing corners of a rectangular room that is 12 feet wide and 16 feet long is 20 feet. Since $12^2 + 16^2 = 400 = 20^2$ deduction allows us to conclude that the distance is 20 feet.

Computer scientists use deduction in many ways. For example, a common rule in algorithms is that for certain problem solutions it is critical that one task occur *before* a second task; this idea of one thing following another is called a *sequential* form of execution because the tasks must occur in sequence. A software developer applies knowledge of this sequencing when encountering a problem such as the software to send an e-mail message. The task of sending a message must be preceded by the task of specifying the recipient's address. A valid algorithm must ensure such task ordering.

For example, consider the integer variables itemCost and priceWith-Tax, and an algorithm consisting of the following sequence of three instructions:

```
priceWithTax ← 0
itemCost ← 100
priceWithTax ← itemCost + itemCost *.055
```

Executing these three instructions in sequence, that is to say from first to last and one at a time, results in the value 105.5 assigned to price-WithTax. However, rearranging the order of the same three instructions changes the algorithm. For example, consider the following algorithm, using the same three instructions:

```
itemCost ← 100
priceWithTax ← itemCost + itemCost *.055
priceWithTax ← 0
```

When this second algorithm executes, the resulting value assigned to priceWithTax is 0. Such subtleties point out the importance of understanding even something as simple as the order of a sequence.

Often the "rules" used by computer scientists are more like *patterns*. As a first illustration of an algorithmic pattern, consider the problem of swapping the content of two variables. Figure 4.5 contains a pattern for a swap algorithm that swaps (interchanges) the content of two variables—varA and varB—using a third variable, temp.

To illustrate how a programmer can use this pattern as a kind of rule, suppose we have two variables: myDog and yourDog. Let's further assume that myDog stores the name "Fido" initially, while yourDog stores "Rover." Now suppose we want to swap these variables so they store the opposite names. We can use the pattern from Figure 4.5 as a general rule and deduce a correct algorithm by substituting variable names (myDog is substituted for varA and yourDog for varB). This substitution results in the following algorithm:

```
temp ← varA
varA ← varB
varB ← temp
```

FIGURE 4.5 The swapping pattern.

```
temp ← myDog
myDog ← yourDog
yourDog ← temp
```

Lest you think that patterns are unimportant. Think what would happen if programmers did not know about the swap pattern and incorrectly assumed that the following algorithm would work to swap myDog and yourDog:

```
myDog←yourDog
yourDog ← myDog
```

Swapping illustrates one kind of algorithmic pattern. Another, more general, pattern of software execution is known as *repetition*. Repetition is used anytime a task needs to be repeatedly performed. Repetition is common in computer programs because one of the things that the computer can do well is to repeat some process, no matter how mundane, over and over. Software developers might recognize a pattern of repeating something five times and apply this pattern to selecting a basketball team or to dealing a five-card poker hand.

Patterns are not restricted to forms of program execution. There are patterns for successful software testing, patterns for ways to provide better software security, patterns of different software performance. Indeed, most of this book can be viewed as an examination of foundational patterns of computation, including the patterns of logic found in Chapter 2.

Another way to think about deduction is to consider it a form of specialization. In deduction we apply our knowledge of more general rules, theorems, or patterns in order to solve a specific problem. So deduction works from the general to the specific. But it is equally important for a computer scientist to be able to work from the specific to the general; this is another application of logic, known as *inductive reasoning*.

Inductive reasoning occurs during the process of recognizing the generalities. For example, many software designs use the idea of separating an algorithm into three components:

1. Software for communicating with the user

2. Software for retrieving and storing data

3. Software for calculating results based on user input and/or retrieved data

This pattern is so common that it is known as a *3-tiered software architecture*. How did we discover this pattern? Such patterns are usually noticed by astute observation that different successful solutions actually use the same general framework

Why is it important to begin to think in terms of 3-tiered software? The answer to this question lies in part with modern computer hardware divisions. Typically, you use your computer to perform a task such as change your personal data on a social networking site. You are communicating with your personal computer (Tier 1), while the changes must occur on the computer owned by the social network company (Tier 3). For efficiency and other reasons, large computer installations store most data on separate database computers (Tier 2). Since three different computers are likely to be involved in processing your social network pages, it makes sense that the software be subdivided into three parts or tiers.

Knowledge of the patterns used in computer science is not only important to computer scientists. For example, knowing about 3-tiered architectures is helpful in understanding that performance of a computer application depends upon the performance of three different systems; any one of these systems could be a bottleneck. Similarly, the 3-tiered architecture has significant implications for data confidentiality.

4.3 DECOMPOSITION: SOFTWARE DESIGN

When software engineers spend time seeking out cause-effect relationships, as discussed in the previous section, they are also using another common problem solving strategy: decomposition. They are decomposing the problem into individual relationships.

Decomposition techniques are often called *divide and conquer* strategies. The idea is to approach a single problem by separating (dividing) it into its constituent subproblems, and then to solve (conquer) each subproblem individually, as depicted in Figure 4.6. Often the subproblems can also be considered as individual problems, and the strategy can be repeated to divide and conquer them.

As an example of this divide and conquer strategy, consider the problem of making a pizza, depicted in Figure 4.7. Notice that the larger problem of making a pizza can be thought of as a collection of six subproblems: (1) making a crust, (2) making and spreading sauce, (3) spreading cheese, (4) spreading toppings, (5) baking, and (6) slicing. A divide and conquer strategy could begin by identifying these six subproblems, then solving the larger problem by solving each of the individual smaller subproblems.

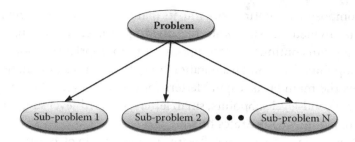

FIGURE 4.6   Divide and conquer means to separate a problem into subproblems.

Computer scientists did not discover the concept of divide and conquer. Evidence of this idea is found in the oft-quoted Aesop's fable regarding the bundle of sticks. This fable dating back to ancient Greece, around 600 BC, tells the story of a father who teaches his sons a lesson. The father asks each son to break a large bundle of sticks that are bound together with string. When the sons discover they are not strong enough to break the bundle as a unit, the father tells them to untie the bundle and break the sticks individually, which they are able to do. The moral of the fable is that many times it is easier to solve a large task by separating it into smaller tasks and solving the smaller tasks, by dividing and conquering.

Some of the reasoning behind the wisdom of divide and conquer was explained in a paper written in 1956 by a psychologist named George Miller [2]. The essence of Miller's work is that the human brain is rather limited in its reasoning abilities and that for some kinds of thinking the human memory is limited to 7 ± 2 items. In practical terms this means that many problems are too complex for the human mind, unless they are approached in terms of simpler partial problems.

Decomposition is frequently taught as a technique for writing, known as *outlining*. The basic notion of outlining is to organize a work beginning

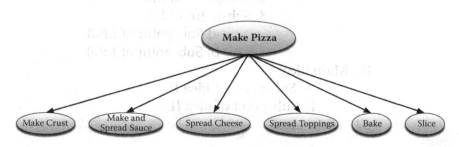

FIGURE 4.7   Decomposing the problem of making a pizza.

by decomposing the entire work into its main ideas. Each of the main tasks might be divided into subpoints, like dividing problems into subproblems. Outlining can continue by decomposing any appropriate subpoint into its own subpoints. The standard notation for outlining uses Roman numerals to index the main tasks, capital letters for the first-level subpoints, numbers for second-level subpoints, small letters for third-level subpoints, and so forth. Figure 4.8 illustrates this.

Section 4.1 explained that software engineering involves three primary activities: analysis, design, and implementation. As described earlier, analysis seeks to define the problem. The implementation activity seeks to create the final software. *Design* work has always been a bit more difficult to explain. In essence the design work is any part of software development intended to progress from the problem definition, or part of the definition, toward but not including writing the final code. Sometimes the design process creates parts of the program; sometimes it results in diagrams. Frequently, the goal of design is to produce an algorithm that can be translated into the final program.

An *algorithm* is defined to be a collection of instructions for performing some task. Computer programs are clearly algorithms, but not all algorithms can be executed by computers. For example, consider cookie recipes. If you search the Internet for the recipe for your favorite cookie, you will discover many algorithms; a good cookie recipe (algorithm) is a

    I. Main idea
           A. Sub-point of idea I.
                1. Sub-point of I.A.
                2. Sub-point of I.A.
           B. Sub-point of I
                1. Sub-point of I.B.
                2. Sub-point of I.B.
                3. Sub-point of I.B.
                      a) Sub-point of I.B.3.
                      b) Sub-point of I.B.3.
   II. Main idea
           A. Sub-point of idea II.
           B. Sub-point of idea II.
  III. Main idea

FIGURE 4.8 Standard outline form.

collection of explicit instructions for how to mix and bake these sweets. The directions for assembling a child's tricycle is another example of an algorithm. These directions can be a combination of English and pictures intended to provide the necessary instructions for solving the assembly task. Neither the recipe for your favorite cookie nor the assembly instructions for a tricycle can be executed directly by a computer; they are not computer programs. However, both are valid algorithms, and both are valuable because they use English or diagrams to convey the task.

One design technique used by software developers is called *top-down design*. Top-down design starts with a summary of the problem and proceeds by successively refining vague instructions into more explicit instructions. For example, consider designing an algorithm to draw the image of an off-road vehicle shown in Figure 4.9.

A top-level design is a refinement (decomposition) of the problem into an algorithm of a few instructions. These top-level instructions are necessarily extremely vague. For the off-road vehicle problem a suitable top-level design would be something like the following five-instruction algorithm:

1. Draw a green grille.

2. Draw bumper just below the grille.

3. Draw the tires behind the grille and bumper.

4. Draw the windshield just above the grille.

5. Draw two auxiliary lights on top of the windshield.

FIGURE 4.9   An off-road vehicle.

1. Draw a green grille.
   1.1. Draw a green grille background.
   1.2. Draw the left headlight.
   1.3. Draw four equally spaced black vertical rectangles for grille openings.
   1.4. Draw the right headlight.
2. Draw bumper just below the grille.
3. Draw the tires behind the grille and bumper.
   3.1. Draw the left tire (a black rectangle) slightly protruding left of the grille and bumper.
   3.2. Draw a half-moon hubcap extending outside the left tire.
   3.3. Draw the right tire (a black rectangle) slightly protruding right of the grille and bumper.
   3.4. Draw a half-moon hubcap extending outside the right tire.
4. Draw the windshield just above the grille.
   4.1. Draw a gray rectangle for the outside of the windshield.
   4.2. Draw a white rectangle for the center of the windshield.
5. Draw two auxiliary lights on top of the windshield.
   5.1. Draw the left auxiliary light.
   5.2. Draw the right auxiliary light.

FIGURE 4.10   Second-level algorithm for the off-road vehicle drawing.

The second step in top-down design is to consider each of the instructions from the top level and one by one refine each instruction into parts. In our off-road vehicle example, drawing the grill involves headlights and black grill openings that can be included to provide more detail. Similarly, the tires, windshield, and auxiliary lights can be further refined. Figure 4.10 shows a suitable second-level algorithm.

Notice that top-down design produces an algorithm that is very much like an outline. Instructions from higher levels are refined into more detailed instructions that are indented beneath. We have chosen to number the algorithm parts a little differently than standard outline numbering in order to clarify the relationship between an instruction and its refinement. For example, Instruction 1 is refined into four more detailed instructions: 1.1 through 1.4.

The process of top-down design does not necessarily end with a second-level design. Anytime the designer feels it is helpful to provide more detail some or all instructions can be further refined. For the off-road vehicle example it seems useful to provide one more level of refinement as shown in Figure 4.11.

1. Draw a green grille.
    1.1 Draw a green grille background.
    1.2 Draw the left headlight.
        1.2.1. Draw black outer dot as a rim for the left headlight.
        1.2.2. Draw white dot centered within the black rim.
        1.2.3. Draw three equally spaced vertical black rectangles as headlight protectors.
    1.3 Draw four equally spaced black vertical rectangles for grille openings.
    1.4 Draw the right headlight.
        1.4.1. Draw black outer dot as a rim for the right headlight.
        1.4.2. Draw white dot centered within the black rim.
        1.4.3. Draw three equally spaced vertical black rectangles as headlight protectors.
2. Draw bumper just below the grille.
3. Draw the tires behind the grille and bumper.
    3.1. Draw the left tire (a black rectangle) slightly protruding left of the grille and bumper.
    3.2. Draw a half-moon hubcap extending outside the left tire.
    3.3. Draw the right tire (a black rectangle) slightly protruding right of the grille and bumper.
    3.4. Draw a half-moon hubcap extending outside the right tire.
4. Draw the windshield just above the grille.
    4.1. Draw a gray rectangle for the outside of the windshield.
    4.2. Draw a white rectangle for the center of the windshield.
5. Draw two auxiliary lights on top of the windshield.
    5.1. Draw the left auxiliary light.
        5.1.1. Draw outer black rectangle as a rim for the left auxiliary light.
        5.1.2. Draw yellow rectangle centered inside the black rim.
        5.1.3. Place the word "HI" centered in the yellow rectangle.
        5.1.4. Draw a small black rectangle for the base of the light.
    5.2. Draw the right auxiliary light.
        5.2.1. Draw outer black rectangle as a rim for the left auxiliary light.
        5.2.2. Draw yellow rectangle centered inside the black rim.
        5.2.3. Place the word "MOM" centered in the yellow rectangle.
        5.2.4. Draw a small black rectangle for the base of the light.

FIGURE 4.11　Final algorithm for the off-road vehicle drawing.

Top-down design is not the only design technique based upon decomposition. Another form of software design that is widely used by software engineers today is known as *successive prototyping*. A *prototype* of any object is an early approximation of the object. For example, automobile manufacturers often create clay or plastic models of potential new cars long before they actually intend to manufacture them. The model (a prototype) allows the company to investigate customer opinions and possibly even perform initial tests for aerodynamics.

A prototype for a computer program is a partially functioning piece of software. Typically, a prototype performs some, but not all, of the requirements from the problem definition. For example, the prototype may display the proper application window, but when the user clicks buttons nothing happens because the functionality associated with the buttons is not included in the prototype.

Designing with successive prototyping consists of a sequence of individual prototypes that progressively lead to a final working program. The early prototypes provide very little of the desired functionality, while later prototypes are nearly finished programs. The design for a new car might begin with a sketch as a first prototype, then proceed to a clay model for a second prototype, followed by concept cars serving as later prototypes.

As an example of successive prototyping of software consider that Sal's Surfboard Shop (SSS) is a retail store, and that Sal wants to expand the business by creating an online store to sell his surfing products. An initial prototype of the SSS online purchase software might produce a homepage that looks something like Figure 4.12.

For this first prototype we will assume that all of the links on this homepage are nonfunctional. Even lacking all of this functionality, this first prototype is useful in the sense that the customer can examine the look and feel of the homepage, perhaps requesting different colors, images, or links. The designer could then produce a second prototype similar to the first but incorporating Sal's requests.

A reasonable succession of prototypes might continue as indicated next:

Prototype 1—Initial web page only (see Figure 4.12)

Prototype 2—Same as Prototype 1 except incorporating Sal's requested improvements

Prototype 3—Same as Prototype 2 except including all inventory pages (no ability to modify cart)

FIGURE 4.12   Initial page for SSS online purchase website.

Prototype 4—Same as Prototype 3 except including ability to add items and view cart

Prototype 5—Same as Prototype 4 except including secure checkout functionality

Prototype 6—Same as Prototype 5 except supporting site search functionality and "Contact us" page

In this case the last prototype (Prototype 6) may provide all required functionality; this prototype could be ready to submit as a final product after sufficient software testing.

Successive prototyping is a form of decomposition, because each prototype represents a focus on some portion of the entire set of requirements. This means that the software engineer has to decompose the entire problem into parts and implement the parts successively.

Both top-down design and successive prototyping have their own advantages. Top-down design can work well when all of the software requirements are all well known from the outset. Top-down design is advantageous because at each step the algorithm is complete, although possibly quite vague. Such completeness provides better confidence that no functionality will be overlooked.

The greatest shortcoming of top-down design is that generally there is no computer software produced until the design is complete. Top-down design documents are algorithms, but these algorithms are not executable. This lack of any kind of actual software throughout the design process can appear to the customer as a lack of progress, as well as making it difficult for project managers to estimate the amount of progress that has been made. Software developers also tend to be more satisfied with the visible evidence of progress provided by actual computer programs. Successive prototyping is popular to a great extent because of the tangible progress evident in each prototype.

When requirements are not completely understood initially, successive prototyping also has a clear advantage. The completion of each prototype provides an opportunity for those involved in the project to evaluate the software and make significant changes to the overall product. Customers in particular like to create new requirements or make changes to requirements in response to a prototype. This customer input has become so integral to software development that the term *content expert* is now used to refer to a person who understands the customer's wishes. Typically, the content expert is not a computer scientist, but still plays a vital role in software development.

4.4 DECOMPOSITION: OTHER USES

Decomposition is certainly an essential skill for software design, but design is far from the only way that decomposition is used in computing. Modern computers often contain multiple processors, also referred to as *cores*. These multicore machines allow for different programs, or different parts of the same program, to execute simultaneously. The idea that multiple pieces of software execute simultaneously, also known as *multitasking*, requires that the software be decomposed. This multitasking is restricted to a handful of cores in most laptop computers but extends to tens of thousands of processors in today's supercomputers.

A related concept, called *grid computing*, makes use of the Internet. Grid computing uses many networked computers to attack the same problem. A computer program decomposes the problem in such a way that each computer can work on solving one part of the problem at the same time that other computers solve other parts.

The kind of decomposition described Section 4.3, like that of multitasking involves dividing a program into separate instructions or groups of instructions. Data decomposition is another use of the concept of divide and conquer. In fact, data decomposition is so important that there is an entire field of computing known as *data organization*.

As an example of decomposing data, we examine two *search algorithms*. A search algorithm is a method for examining a group of data items in order to find an item with some particular property.

The most intuitive of all search algorithms is *linear search*. A linear search requires that the group of items be arranged one after another from first to last. The algorithm consists of examining the first item, then the second, the third, the fourth, and so on, until the desired item has been found.

As an example of a linear search, suppose you were given a stack of award certificates and told that one of the awards might be awarded to you. A linear search would proceed as follows looking for your certificate in a stack of 500:

Step 1   Check the name on the top certificate, if not your name then proceed to Step 2.

Step 2   Check the name on the 2nd certificate, if not your name then proceed to Step 3.

Step 3   Check the name on the 3rd certificate, if not your name then proceed to Step 4.

...

Step 500   Check the name on the 500th certificate.

If you find your name on a certificate, you can stop at that step. In the worst case (when your certificate is last or not included) you would need to examine all 500 certificates.

Now suppose that the stack of award certificates is sorted alphabetically with the beginning of the alphabet at the top of the stack. In this case we could use a different algorithm known as *binary search*. Each step of a binary search examines the middle item of the remaining group. By comparing the middle item to the sought item, half of the group can be eliminated from further consideration in this search. To see why consider that your name is Jones, Susan and the name in the middle is Smith, John. Since the stack has been alphabetized and since Jones precedes Smith alphabetically, there is no need to consider the half of the data from Smith to the end. Similarly, if the middle name was Gomez, Marc, then the first half of the data (through Gomez) can be eliminated from further consideration. A binary search of the previous (alphabetized) stack of 500 scholarship certificates would proceed as follows:

Step 1   Check the name on the middle certificate in the stack. If the middle certificate alphabetically precedes your name, then set aside the top half of the stack. If the middle certificate alphabetically follows your name, then set aside the bottom half of the stack.

Step 2   Check the name on the middle certificate in the remaining stack. If the middle certificate alphabetically precedes your name, then set aside the top half of the remaining stack. If the middle certificate alphabetically follows your name, then set aside the bottom half of the remaining stack.

Repeat Step 2 until either your certificate is found or the remaining stack has only one certificate that is not yours.

As mentioned earlier, a disadvantage of binary search is that it requires data that is sorted. However, the payoff is that binary search is usually faster. A linear search of 500 scholarship certificates can require as many as 500 steps, while the most steps for binary search is 9. Figure 4.13 justifies this claim.

Both linear search and binary search involve decomposition of data in the sense that they divide the data into individual parts to be searched, but the way that this division occurs is quite different. Linear search removes one item from the remaining group of items for each step of the algorithm, whereas binary search removes roughly half of the group at each step. This distinction in the way that divide and conquer is applied leads to the significantly improved performance of binary search. For more discussion of how decomposition is useful for data, Chapter 7 considers more data organizational techniques.

4.5 ABSTRACTION: CLASS DIAGRAMS

Sometimes the word "abstract" conjures up thoughts of abstract art that might seem unusual or even extreme or difficult to understand. However, there is another definition for the technique used by computer science. A more appropriate definition is that an *abstraction* is anything that allows us to concentrate on important characteristics while deemphasizing less important, perhaps distracting, details. For example, you are

Binary Search	
after this step	number of items remaining to be searched
1	250
2	125
3	62
4	31
5	15
6	7
7	3
8	1
9	search complete

Linear Search	
after this step	number of items remaining to be searched
1	499
2	498
3	497
4	496
5	495
6	494
7	493
8	492
9	491
10	490
...	...
499	1
500	search complete

FIGURE 4.13   Worst-case performance of linear and binary search (500 items).

using abstraction if you tell someone that you just saw a red Corvette. In this case the color and model of the car were considered to be the most important, while details of the model year, engine displacement, wheel dimensions, and so forth are omitted. Abstraction allows for more succinct communication, while also allowing the user to choose a particular emphasis. Someone else might have felt that the year the Corvette is manufactured is more important than its color and use a different abstraction.

Software engineers use abstractions that assist in the problem-solving process. In Section 4.3 we pointed out human memory limitations of 7 ± 2. One way to overcome our inability to solve complex problems is to simplify complexity using abstraction. Like a digital camera uses a handful of focus points, computer scientists learn to focus on the most important issues through abstraction.

A *control structure* is a mechanism for specifying the proper order in which instructions must be performed. Algorithms are composed of five fundamental control structures:

1. Sequential control

2. Selection

3. Repetition

4. Control abstraction

5. Concurrency (the topic of Chapter 11)

These five structures are composed to define the *control* flow—the order of instruction execution—for an algorithm. The first three of these control mechanisms were introduced in Section 4.2. Sequential control ensures that one instruction executes before another. Selection occurs whenever the algorithm must choose (select) among different options. Repetition consists of executing one or more instructions repeatedly. As an example of these three control structures, consider the recipe (algorithm) for making fudge in Figure 4.14.

The fudge recipe demonstrates many examples of sequential execution. Each of the numbered steps necessarily precedes the subsequent steps in the sequence of execution. Control selection occurs in Step 2, when the cook has the option of adding walnuts or not. Step 5 is an example of control repetition.

Control abstraction is not necessary in an algorithm as simple as the fudge recipe. However, if we consider something more complicated, then control abstraction becomes useful. Figure 4.15 illustrates with directions (an algorithm) for creating a tray of treats.

Control abstraction occurs when one instruction in an algorithm consists of a reference to executing a subalgorithm. In the algorithm from Figure 4.15, Step 1 is such a reference. This single instruction abstractly refers to the recipe (a kind of subalgorithm) from Figure 4.14. Notice that

RECIPE STEPS

1. Place the following ingredients in a microwavable bowl:
   3 cups of semisweet chocolate chips
   one 14 oz. can of sweetened condensed milk
   ¼ cup of butter
2. If desired, stir in 1 cup of walnut pieces.
3. Zap in microwave for one minute.
4. Remove from microwave and stir the mixture.
5. Repeat Steps 3 and 4 until chocolate chips are completely melted.
6. Stir 1 teaspoon of vanilla into mixture.
7. Pour mixture into a greased 8 by 8 dish.
8. Refrigerate for three hours.
9. Cut fudge into 1 inch squares.

FIGURE 4.14   Fudge recipe.

control abstraction removes distracting details from the algorithm for making the tray of treats. Step 1 abstracts the nine-step recipe for fudge in just two words: Make fudge. Similarly, Steps 2 and 3 are probably also abstractions for other cooking recipes.

Data abstraction is no less important than control abstraction. A standard way to diagram data abstraction is known as a *class diagram*. Class diagrams use a rectangle to denote a single class of objects. The class diagram rectangle abstracts the group of objects in terms of two things:

- *Attributes*

- *Operations*

The theory behind this kind of data abstraction is that objects in our world can be explained in terms of (1) what they are and (2) what they can do. An object's attributes capture what it is, and operations that can be performed upon the object define what it can do. For example, consider the class diagram for a thermostat from Figure 4.16.

RECIPE STEPS

1. Make fudge (see Figure 4.14).
2. Make chocolate chip cookies.
3. Make peanut butter bars.
4. Arrange fudge, cookies, and bars on a large tray.

FIGURE 4.15   Making a tray of treats.

Thermostat
heatSwitchSetting : (COOL / OFF / HEAT) fanSetting : (ON / AUTO) temperatureSetting : integer
setToHeat () setToCool () setToNoHeat () setFanToOn () setFanToAuto () increaseTempSetting () decreaseTempSetting () readCurrentTemperature ()

FIGURE 4.16 Thermostat class diagram.

The rectangle on the left abstractly describes the thermostat pictured to its right. A class diagram rectangle consists of three parts, diagrammed in three horizontal compartments. The name of the class of objects (in this case *Thermostat*) is in the top compartment. The middle compartment lists attributes. The thermostat can be abstracted into three attributes: (1) the current position of the upper left switch (COOL, OFF, or HEAT); (2) the current position of the upper-right fan switch (either ON or AUTO); (3) the current setting of the rotary temperature dial.

The operations in a class diagram are listed in the bottom compartment. Operations are abstract references to the behavior of the object. There are three ways to change the upper-left switch of the thermostat, corresponding to the setToHeat, setToCool, and setToNoHeat operations. Similarly, there are two ways some can position the upper-right switch: setFanToOn and setFanToAuto. Turning the temperature dial clockwise is listed as increaseTempSetting and turning counterclockwise as decreaseTempSetting. The last operation, readCurrentTemperature, is included because homeowners often use thermostats to check the house temperature. In total this kind of thermostat is abstracted as three attributes and eight operations that capture the most important aspects of the device.

The pair of parentheses following each operation in Figure 4.16 is indicative of the potential for operations to include parameters. A parameter provides flexibility to a method by allowing the same method to receive different argument values. In the case of the Thermostat, we could have

Thermostat Too
heatSwitchSetting: (COOL / OFF / HEAT)   fanSetting: (ON / AUTO)   temperatureSetting: integer
setMainFunction(f : COOL / OFF / HEAT)   setFan (b : ON / AUTO)   setTemperature (t : integer)

FIGURE 4.17   ThermostatToo class diagram.

replaced the increaseTempSetting and decreaseTempSetting methods by the following single method:

```
setTemperature(t : integer)
```

The setTemperature method has a single parameter, named t. The ": integer" notation indicates that t must be some integer. When the set-Temperature is applied to a thermostat, an integer argument must be supplied. For example, setTemperature(70) represents an operation to set the temperature to 70 degrees.

Figure 4.17 shows a second Thermostat class that incorporates parameters in its operations. This class requires fewer operations because of the flexibility provided by parameters. In place of applying a setToNo-Heat() operation, the equivalent operation for ThermostatToo objects is setMainFunction(OFF). Similarly, the operations setMainFunction(COOL), setMainFunction(HEAT), setFan(ON), setFan(AUTO) from ThermostatToo correspond, respectively, to setToCool(), setToHeat(), setFanToOn(), and setFanToAuto().

As a second example class diagram, consider a dial padlock (see Figure 4.18). The padlock has two attributes (dialPosition and latch) and four operations, as shown.

4.6 ABSTRACTION: USE CASE DIAGRAMS

We have seen that abstraction is useful during software design both for expressing the instructions of an algorithm and for describing algorithm's data. But abstraction can be used even before the process of software design begins.

Padlock
dialPosition : integer   latch : (OPEN / CLOSED)
turnLeftTo(j : integer)   turnRightTo (j : integer)   pullLatchOpen()   pushLatchClosed()

FIGURE 4.18   Padlock class diagram.

Use case diagrams are a technique for depicting a system—software system or some other system—by way of interaction between computer users and a system. The two main components of a use case diagram are *actors* and *use cases*. Each actor represents a group of users of similar type, and a use case is an action that can be performed by the system. Lines are drawn from each actor to the particular actions that this type of user can perform.

Figure 4.19 is a simple use case diagram that describes the work of a cab driver. The single actor shown in this diagram is the cab driver and there are four use cases to abstract four significant actions performed by the actor.

Figure 4.20 is a use case diagram for a more complex system: a computer program for online student course registration. In this case a rectangle has been drawn around the actions to indicate that they collectively represent the functionality of the registration system.

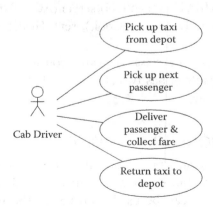

FIGURE 4.19   Cab driver use case diagram.

Registration System

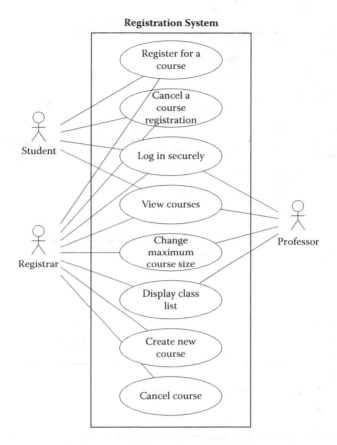

FIGURE 4.20   Registration system use case diagram.

Actors of a use case diagram are grouped into *roles*. The reason for the different roles is to group users according to their shared actions. In the registration example there are three roles and, therefore, three actors: student, registrar, and professor. Students can perform four actions: registering for a class, canceling a class registration, logging in, and viewing online courses. Professors do not need to register for classes or cancel such enrollment, but they might want to view their class lists or increase/decrease the number of possible students who can enroll. The registrar needs to be able to perform still other actions. It is common for some actions to be available to multiple actors. For example, students, professors, and registrars must all be able to log into the student registration system.

Use case diagrams may also depict relationships between actions. Two of the more common relationships are labeled «extend» and «include».

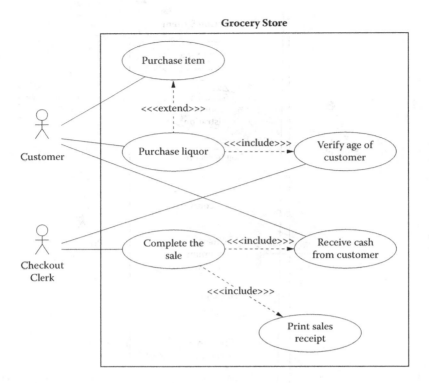

FIGURE 4.21   Grocery store use case diagram.

An «*extend*» relationship occurs whenever one action is an extension or specialized version of another. An «*include*» relationship results from one action making use of another as part of its function. Figure 4.21 contains examples of both of these relationships.

This use case diagram illustrates a simplified version of the checkout system in a grocery store. Two actors—the customer and the checkout clerk—are pictured. The primary action of a customer is to purchase items from the store. However, liquor is handled somewhat differently because of the need to be certain that the customer is of legal age to purchase liquor. Therefore, the action of purchasing liquor is a specialized version of making a purchase.

There are three «include» relationships pictured in Figure 4.21. The action of purchasing liquor involves the action of verifying the customer's age. Likewise, receiving money from the customer and printing a sales receipt are both actions that are a part of completing the sale.

4.7 SUMMARY

The four problem-solving techniques—problem definition, logical reasoning, decomposition, and abstraction—that make up the foundation of this chapter form an interconnected structure that are at the core of computational thinking. Problem definitions are decomposed into functional requirements that are typically expressed abstractly using forms borrowed from logic. Deductive reasoning is based in logical reasoning. The 3-tiered design model is itself a decomposition into three abstract modules, one of which is focused on the logic of the processing solution. Similarly, top-down design, prototyping, and data decomposition all rely upon problem definition, logical reasoning, and abstraction.

The five patterns of control—sequential execution, selection, repetition, control abstraction, and concurrency—are essential for expressing algorithms, and algorithms are the form of problem solutions in the software world. Diagrams are also helpful in the process of designing software solutions. Use case diagrams are used to analyze the problem; state-activity diagrams help ensure completeness of a set of functional requirements; and class diagrams are effective tools for software design.

Most of the problem-solving techniques of this chapter had their origins long before computers existed and most have utility that extends well beyond computer science. What employee could not benefit by improved problem-solving skills? Occupations from scientific research to crime scene investigation are rooted in logical reasoning. Sociologists, economists, and politicians all decompose populations into abstractly defined groups—another form of data decomposition.

4.8 WHEN WILL YOU EVER USE THIS STUFF?

Life can be viewed as an endless sequence of problems. When you wake up you must solve the problem of what clothes to wear and what to eat for breakfast, and at the end of the day you must decide when to retire and when to set the alarm clock for your next day. The middle of our workdays tend to result in more complex problems, and it is these more challenging problems that determine things like success and paycheck size. It is also these more challenging problems that benefit from well-honed problem-solving skills.

How many times have you been overwhelmed by a new project, saying something like, "I don't know where to begin"? There are two good answers to this question in this chapter: (1) begin by defining the problem, and (2) you can always limit the focus of the problem by using abstractions. The chapter goes further to teach strategies, such as listing functional requirements for problem definition and use case diagrams for abstraction.

Many problem-solving strategies are not unique to computer science. Divide and conquer problem solving is something many would label as common sense. However, computer science applies divide and conquer in many ways from top-down design to prototyping, from dividing data in half for a binary search to identifying simultaneous efforts in multitasking.

Logical reasoning also permeates all of humankind's discoveries. From a few experiments, scientists can generalize findings by careful use of inductive reasoning. The scientist might also rely upon the patterns from prior observations as a way to determine which future experiments are of interest, thereby applying a form of deductive reasoning.

Yes, we are all problem solvers. But when it comes to problems with algorithmic solutions, no one understands better how to solve the problems than the computer scientist.

REFERENCES

1. Doyle, Sir Arthur Conan. *The Sign of Four*. (1890).
2. Miller, G. "The magical number seven, plus or minus two: Some limits on our capacity for processing information." *The Psychological Review* 63 (1956): 81–97.

TERMINOLOGY

3-tiered architecture	control abstraction
abstraction	control flow
actor	control structure
algorithm	customer (software)
analysis	data abstraction
attributes (of a class)	data organization
binary search	decomposition
class diagram	deductive reasoning
content expert	design (of software)

divide and conquer	prototype
functional requirements	repetition
grid computing	requirements (software)
inductive reasoning	role (of users)
linear search	search algorithm
logical reasoning	sequential (execution)
multitasking	state-activity table
nonfunctional requirements	successive prototyping
operations (of a class)	top-down design
outlining	use case
patterns	use case diagram
problem definition	

EXERCISES

1. Write a group of functional requirements in the style of Figure 4.1 for the task of setting the time on a common wristwatch.

2. Draw a class diagram for the wristwatch described in the previous exercise.

3. Draw a class diagram for the DVD player described in Figure 4.1.

4. Create a use case diagram for handling books in a library. Be certain to capture the work of patrons and librarians.

5. For each of the following situations, describe whether the reasoning is *deductive* or *inductive*.

   a. Scientific investigation relies upon two different laboratories that are able to repeat the same experiment before publishing the result.

   b. Using the current weather radar and local forecast for the day, you decide whether to schedule an outdoor party.

   c. Based upon the US Constitution lawyers argue a specific case in front of the Supreme Court.

d. A mathematician proves a new theorem by applying accepted axioms and other theorems as reasoning for each step of the proof.

e. A painter decides whether to use oil or water colors for a particular painting based upon past experiences using these different media.

f. A forensic expert concludes that someone died of natural causes, based upon her education and the results of an autopsy.

6. Consider the following set of functional requirements:

Index: W1
Name: Parka
Action: If the current temperature is less than 0 degrees Fahrenheit and the forecast is for snow, then you should wear a parka.

Index: W2
Name: Raincoat
Action: If the current temperature is greater than 0 degrees Fahrenheit and the forecast is for snow, then you should wear a raincoat.

Index: W3
Name: Umbrella
Action: If the current temperature is greater than 0 degrees Fahrenheit and the forecast is for precipitation, then you should carry an umbrella.

Index: W4
Name: Leather
Action: If the current temperature is less than 0 degrees Fahrenheit and the forecast is for no precipitation, then you should wear a leather coat.

a. Create a state-activity table for this set of requirements.

b. Identify all of the conflicts in the set of requirements.

c. Identify all of the ways in which the requirements are incomplete.

7. In Section 4.2 the following algorithm is proposed as an incorrect way to attempt to swap the contents of two variables. Assuming that the myDog variable is assigned "Fido" initially, while yourDog stores "Rover," then what values do each variable store after this algorithm executes?

```
myDog   ← yourDog
yourDog ← myDog
```

8. Suppose you are given the task of looking for a particular name in a telephone book with roughly 4,000 names.

   a. Which name of the phone book would be the first one examined using a linear search?

   b. Which name of the phone book would be the first one examined using a binary search?

   c. Assuming the person you are searching for is not in the phone book, how many names would be examined to discover this when using a linear search?

   d. Assuming the person you are searching for is not in the phone book, about how many names would be examined to discover this when using a binary search?

CHAPTER 5

Algorithmic Thinking

And now the sequence of events in no particular order.

—**DAN RATHER**

OBJECTIVES

- Understand the concept of software and program execution
- Understand that algorithms involve choices and choices take the form of selections involving logical conditions
- Understand that algorithms often involve the repetition of statements
- Understand how algorithms are modularized
- Understand the form and function of flowchart elements for imperative statements including naming, selection, and repetition
- Understand the concepts of computational state, events, and actions
- Model sequential algorithms of ten or fewer states

An algorithmic view of nature, social interaction, and life itself can be incredibly valuable. Ranging from the mundane to the complex, many of life's essential activities involve following a sequence of simple and discrete steps to solve some larger problem. Baking a dozen cookies, tying a Windsor knot, developing a business plan, diagnosing an illness, and designing a commercial jumbo jet are examples of processes that require algorithmic thinking to complete.

Perhaps more than any other discipline, computer science has given deep thought to algorithmic thinking and has developed highly refined ways of describing and interpreting complex processes. This chapter describes actions, sequences of actions, conditional actions, repeated actions, and how to reduce complexity by decomposing large problems into smaller units.

5.1 ALGORITHMS

An *algorithm* is a sequence of discrete actions that when followed, will result in achieving some goal or solving some problem. Everyone is accustomed to following algorithms in daily life whether we are aware of it or not. Football players are following an algorithm when they execute a play on the field; drivers are using an algorithm when they follow a set of instructions for getting from one city to another; piano players are using an algorithm when they read and play a musical score; mathematicians are using an algorithm when they use long division to find the ratio of two numbers; and actors are using an algorithm of sorts when they read and execute a script for some theatrical performance.

As another example, consider that a recipe is an algorithm that a chef can follow to create a main dish, dessert, or even a drink.

Chocolate Chip[1] Cookie Recipe

Ingredients
- 1 cup melted butter
- 2 cups brown sugar
- 2 eggs
- 3 cups flour
- 1 teaspoon baking powder
- 1 teaspoon baking soda
- 2 cups chocolate chips

Directions
1. Preheat the oven to 375 degrees F.
2. Line a cookie sheet with parchment paper
3. In a bowl, stir together the butter, brown sugar and eggs.
4. In a separate bowl, combine the flour, baking powder and baking soda. Gradually combine with the sugar mixture.
5. Add the chocolate chips
6. Fill the cookie sheet with one-spoonful drops of the cookie dough.
7. Bake dough for 9 minutes
8. Cool for five minutes before removing from cookie sheet.

Image courtesy of http://www.flickr.com/photos/amagill/34754258/

In the field of computer science, an *algorithm* is any well-defined sequence of actions that takes a set of values as input and produces some set of values as output. In other words, an algorithm is a sequence of computational actions that transform the input into the desired output. A computational understanding of the chocolate chip cookie recipe views the list of ingredients as the input to the recipe; the directions constitute the algorithm. The directions describe an orderly sequence of discrete actions that transform the input into the desired output. The output is, of course, a plate full of the best chocolate chip cookies in the world.

Any useful algorithm must ensure that several important properties are maintained. One of the most important requirements is that each of the individual actions referred to in the algorithm be meaningful. When computer programmers write an algorithm, programmers must know the precise meaning of each action they write; just as a chefs must understand the meaning of each action described in a recipe. Computer programmers refer to the meaning of an action as the *semantics* of an action. The phrase *semantics* refers to the meaning of the actions that occur in an algorithm.

Another requirement is that the actions of an algorithm must have only one possible interpretation, so that there is no misunderstanding about the semantics of the action. We say that an action must be *unambiguous*; meaning that the action is not subject to conflicting interpretations. In the recipe example, each step is written in such a way that a chef can easily understand and therefore follow the instructions. Actions like *preheat*, *line*, *stir*, *combine*, *add*, *bake*, and *cool* are common terms that all chefs fully understand and are not subject to multiple interpretations. If we replace the word *bake* in step 7 with the term *heat*, the directions now say to "*Heat* dough for 9 minutes." This makes the algorithm less clear and therefore more difficult to follow since chefs do not typically use this term to describe baking cookies. An inexperienced cook might consider broiling the dough or simply letting it set out at room temperature in order to heat the dough.

An algorithm must also ensure that the order in which the actions occur be well defined. It should be self-evident that the order of actions taken when making chocolate chip cookies is vitally important. If, for example, we swap steps 6 and 7 in the cookie recipe, we end up placing a bowl full of dough into the oven and baking one giant cookie that we remove from the oven and carve up into small spoon-sized chunks that we carefully drop onto the cookie sheet after which the dough is allowed to bake for nine minutes.

Finally we note that the number of actions described by an algorithm must be finite rather than infinite. No goal can be reached if we are required to follow a never-ending sequence of actions in order to achieve the goal. A computer scientist might express this requirement by saying that an algorithm must *halt* in order to be useful.

5.2 SOFTWARE AND PROGRAMMING LANGUAGES

A computer is able to perform a surprisingly small number of relatively simple actions such as adding two integer numbers or checking to see if two integer numbers are equal. Of course, computers can solve tremendously complex problems and achieve incredibly useful outcomes if they perform these actions quickly enough and in the right sequence.

Computer software, also referred to as a *program*, provides the instructions for telling a computer the algorithm that it should follow to achieve some goal. Software cannot be expressed in human languages such as English or Chinese or Spanish since these languages are not easily used to describe computation with the necessary technical precision. Computer software is therefore expressed in a *programming language*. A programming language is a language that is designed to precisely and compactly express computational algorithms. Programming languages must therefore be able to represent the input and output data of an algorithm along with the sequence of actions that a computer must follow to transform the input into the desired output. When a computer *executes* a computer program, the computer simply follows the steps that are given by the program in order to obtain the desired output from the provided input.

Just like there are numerous human languages in the world, there are a surprisingly large number of different programming languages in common use today. Although each programming language must be able to precisely express a computational algorithm, not all programming languages express computation in the same way. Many programming languages express computation as a sequence of commands that are issued to the computer. These commands are often referred to as imperatives and languages that express computation in this fashion are therefore known as *imperative* languages. Other languages express computation using very different paradigms. *Functional* languages emphasize the relationship between input and output as an intrinsic entity rather than viewing computation as a sequence of individual actions that move from input to output. *Declarative* languages emphasize the expression of facts from which output can be generated by applying relevant facts to the input. Other paradigms include logical programming, object-oriented programming, and concurrent programming.

In this chapter we will use an informal imperative language for expressing computational algorithm. Although a computer cannot directly execute this informal language, the programs that we will describe can be easily translated into a real programming language.

5.3 ACTIONS

As we have already mentioned, computational algorithms must be expressed as a well-defined sequence of actions that the computer must follow. Perhaps the most critical aspect of computation involves knowing what actions a computer can take and also knowing what actions a computer cannot take! Consider, for example, the algorithm shown in Figure 5.1 that gives actions allowing us to travel from New York City to Los Angeles without consuming any fuel or electrical energy. Although the algorithm is very simple to write and very simple to understand, it is not useful since it requires us to perform an impossible action. The action requiring a person to jump 2,440 miles is of course impossible to perform and this renders the algorithm meaningless.

Since any executable algorithm must be expressed as a sequence of *possible* actions, we must first understand what actions a computer might possibly take. Such actions are sometimes referred to as *computable* actions or, more formally, as *computable functions*. The remainder of this section informally defines a small but powerful set of computable functions that allow us to express a surprisingly large number of very useful algorithms.

5.3.1 Name Binding

One of the simplest, yet also one of the most important, actions that a computer can perform is that of naming the elements of a program. In

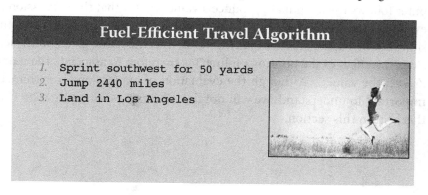

FIGURE 5.1 Unrealizable algorithm for fuel-efficient travel.

Name Binding
IDENTIFIER ← EXPRESSION

FIGURE 5.2 How to write a name binding. The value on the right side of the arrow is bound to the identifier on the left of the arrow.

programming languages, *name binding* is the association of a name, also known as an *identifier*, with a value. Once an identifier is bound to a value, the identifier is said to reference the corresponding value such that when the identifier is used in some later context, the identifier stands for the value to which it is bound.

We will express the name binding action of a computer using an arrow symbol as shown in Figure 5.2. A valid identifier must appear on the left of the arrow and a value or a formula that produces a value must occur on the right. Although most programming languages allow identifiers to contain characters like dashes, underscores, and even dollar signs, we will assume that an identifier must only contain the letters A through Z in either upper- or lowercase. An element known as an *expression* must appear on the right of the arrow. An expression is either a value or a formula that produces a value.

Consider the following sequence of name bindings. The first two actions associate the name X with the value 3 and the name Y with the value 4. Note that in the final expression, the phrase "3 + 4" is an expression that produces the value 7. In computer science, an expression is a short formula that describes how to compute a value. Whenever a computer follows the formula to produce a value, we say that the expression is evaluated. Name binding will always evaluate the expression on the right side and then associate the resulting value with the name given on the left side. The effect of the binding on line 3 is therefore to associate the name Z with the value 7. Although the evaluation of expressions is extremely important to understand, we will not describe expression evaluation further within this section.

1. X ← 3

2. Y ← 4

3. Z ← 3 + 4

Once a binding is established, the name can be used as a replacement for the value to which it is bound. Consider, for example, the following sequence of name binding actions. In this example, the first two actions associate the name X with the value 3 and the name Y with the value 4 as before. The final action associates the name Z with the sum of X and Y. Since X refers to 3 and Y refers to 4 we see that Z becomes bound to the value 7.

1. $X \leftarrow 3$

2. $Y \leftarrow 4$

3. $Z \leftarrow X + Y$

5.3.1.1 Proper Naming

Computer scientists have long recognized the critical importance of choosing correct names for the elements of a program. The importance of proper naming is not a modern invention of computation. In ancient literature and philosophy, the ability to properly name things indicated a mastery over those things. The ancient Middle Eastern book of Genesis, for example, describes how God created the universe and placed the first man, Adam, as the caretaker over all other creatures. Adam demonstrated his mastery by giving a name to every other creature.

Computer scientists also recognize that a set of well-named program-ming elements shows a mastery of the logic and nature of the underly-ing computation. An element is well named if the name descriptively and correctly reflects the central essence of the element. Proper naming thus enables a computer scientist to more clearly reason about the algorithm and data of the program. Poorly named program elements reflect a con-fused understanding of the elements and also make clear-headed reason-ing more difficult.

Consider, for example, an algorithm that takes, as input, the number of gallons of gasoline it takes to fill up a car's gas tank and the cost per gallon of gas; a cost that is assumed to be $3.75. The algorithm will then compute the total cost required to fill up the car's gas tank if the tank is empty. Figure 5.3 shows two algorithms that perform the same set of actions, and hence are identical, except that the names of the data values differ.

It should be apparent that the naming shown on the left reflects a deeper and more correct understanding of the problem than the variable nam-ing on the right. The identifier "Cents," for example, is factually incorrect

Two Gas Cost Algorithms		
1. DollarsPerGallon ← 3.75	*1.* Cents ← 3.75	
2. TankCapacity ← 10	*2.* Size ← 10	
3. DollarsToFill ← TankCapacity × DollarsPerGallon	*3.* Money ← Size × Cents	

FIGURE 5.3   Two algorithms that differ only in the way data are named.

since the number 3.75 is not given in cents but is a dollar amount. Also, the identifier "Cents" does not convey the important information that the number 3.75 represents the cost per gallon of gasoline. The identifier "Size," while not truly incorrect, reflects a vague understanding that it is the capacity of the tank that is central to the problem and not, in general, the physical dimensions or actual size of the tank. Finally, although the identifier "Money" actually reflects that the bound value should be interpreted as currency, it does not accurately convey that it is the cost required to fill the gas tank.

Name binding allows us to write algorithms that solve a wide range of problems. Consider any problem that involves converting a quantity from one unit of measure into another. Perhaps we are traveling to the 2016 Summer Olympic Games in Rio de Janeiro, Brazil, and we need to know how many Brazilian reais (BRL: the official Brazilian unit of currency) one thousand U.S. dollars (USD) is worth. Of course, the conversion is dependent upon the prevailing exchange rate and since the exchange rate will vary from day to day, it may be useful to write a computer program to generate the correct answer for us. Figure 5.4 gives an algorithm using an exchange rate of 205.5 to determine the BRL that corresponds with US$1,000.

Convert from USD to BRL Algorithm

1. USD ← 1000
2. ExchangeRate ← 205.5
3. BRL ← USD × ExchangeRate

FIGURE 5.4   Currency conversion algorithm.

Ensuring that all of the variables in a computer program use correct units of measure is a challenging task for programmers. In September 1999 the $327.6 million Mars Climate Orbiter was lost due to a software error. The satellite entered the Martian upper atmosphere at too low an altitude and disintegrated. NASA later determined that the ground station software was working in pounds while the spacecraft expected to receive the same information in newtons. This inconsistency led the ground station to underestimate the power of the spacecraft's thrusters by a factor of about 4.5.

5.3.1.2 State

Name bindings are allowed to change as a program executes. For example, consider a program that executes the following sequence of name bindings.

1. $X \leftarrow 3$

2. $X \leftarrow 4$

3. $X \leftarrow X + X$

After the computer has executed the first action, the name X is associated with the value 3. After the second action has been executed, the name X is associated with the value 4 rather than the value 3. In the third action, the name X becomes associated with the value produced by the expression "X + X." Since X is associated with the value 4 when the expression "X + X" is evaluated, the name X becomes associated with the value 8.

The *computational state* of a program is the collection of name bindings that are active at any single point in time. Since any programming action will potentially change the state of a program, we can describe the meaning of a program by examining the sequence of changes made to the state as the program executes. Consider the following sequence of name bindings where the program state that results from each name binding is described below the action.

1. $X \leftarrow 3$

The state is now {X = 3}

2. $Y \leftarrow X + 4$

The state is now {X = 3 and Y = 7}

3. Y ← X + Y

The state is now {X = 3 and Y = 10}

4. X ← X * Y

The state is now {X = 30 and Y = 10}

To illustrate how state changes can be useful in the context of software programming we will again consider traveling to the 2016 Summer Olympic Games in Rio de Janeiro. Once we are in Rio, we might need to know the weather forecast for the day to decide whether to attend an outdoor event like skeet shooting or an indoor event like high diving. Perhaps this presents a challenge to us since the temperature in Brazil is reported in degrees Celsius but we are far more familiar with temperatures reported in Fahrenheit. To address this challenge we decide to write a conversion algorithm; an algorithm to convert a temperature from Celsius into Fahrenheit.

We research how this conversion is performed and realize that given a value in Celsius, we first multiply that value by 9, then divide by 5 and finally add 32 to obtain the corresponding value in Fahrenheit. We decide to write the steps of this algorithm as shown in Figure 5.5.

Figure 5.6 shows how the computational state of an algorithm changes as the algorithm is carried out. In particular, we note that the values bound to Fahrenheit change three times as the program executes. Fahrenheit first takes the value 301.5, the product of 33.5 and 9. This binding is then changed such that Fahrenheit refers to 60.3. Finally, we add 32 and bind the name Fahrenheit to the value 92.3. The first two values to which the

Convert from Celsius to Fahrenheit Algorithm

```
1.  Celsius     ←  33.5
2.  Fahrenheit  ←  Celsius * 9
3.  Fahrenheit  ←  Fahrenheit / 5
4.  Fahrenheit  ←  Fahrenheit + 32
```

FIGURE 5.5 Temperature conversion algorithm showing how name bindings are allowed to change.

Convert from Degrees Celsius to Fahrenheit
1. Celsius ← 33.5
The state is now {Celsius = 33.5}
2. Fahrenheit ← Celsius * 9
The state is now {Celsius=33.5 and Fahrenheit=301.5}
3. Fahrenheit ← Fahrenheit / 5
The state is now {Celsius=33.5 and Fahrenheit=60.3}
4. Fahrenheit ← Fahrenheit + 32
The state is now {Celsius=33.5 and Fahrenheit=92.3}

FIGURE 5.6 The computational state changes as the algorithm executes.

name is bound can be consider temporary values; values that are helpful as we make progress toward computing the final answer. The binding of the name Fahrenheit at the conclusion of the algorithm, however, must ultimately be the correct and final result that we seek.

5.3.2 Selection

Control flow is a computational term that refers to the specific order in which the individual actions of a computer program are executed. Normally the actions that a program performs are done in the order that they are written by the programmer. Often, however, the ordering of actions must be flexible and allow the computer to respond to inputs of various types. This flexibility is supported in programming languages by elements known as *control flow statements*. Control flow statements are commands that control the specific ordering of the actions performed by a computer program.

A *selection statement* is a control flow statement that allows a computer to make choices regarding whether certain actions should be performed. A *one-way selection* statement allows a programmer to either perform an action or skip the action. A *two-way selection* statement allows the computer to choose one of exactly two actions, whereas *a multiway selection* statement allows the computer to choose one of several alternatives. We will first describe a one-way selection statement and then expand our discussion to two-way and multiway selection statements.

5.3.2.1 One-Way Selection

Consider, for example, an online store that charges a $10 shipping cost for any order under $40 but gives free shipping for any order of at least $40. This policy can be expressed in English as "If the order amount is less than

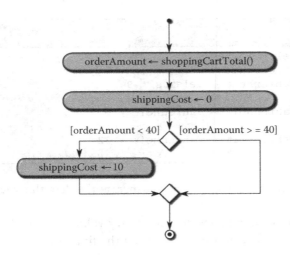

FIGURE 5.7   A flowchart for computing the shipping cost.

$40, then the shipping cost is $10 otherwise the shipping cost is $0." In this example, the shipping cost is set to either $10 or to $0 depending on whether the order amount is less than $40.

We will describe algorithms using two different techniques. First, we will use flowcharts to visually highlight the flow of control through an algorithm. A *flowchart*, also known as an *activity diagram*, is a diagram that helps to visualize the sequence of actions that occur within a complex algorithm. In this chapter we will simply use flowcharts as an intuitive aid to understand sequencing, whereas in Chapter 6 we give a detailed explanation of how flowcharts are used as a high-level model for computational processes. Most computer programs are not written as flowcharts, however, so we will also textually describe algorithms using notations that are very similar to what a computer programmer might use to develop computational algorithms.

Figure 5.7 shows a flowchart that expresses the algorithm for computing the shipping cost for our online shopping site. The circle at the top of the flowchart denotes the starting point of the algorithm and arrows denote the sequencing of the computable actions.

In this flowchart, the order amount, given by the shopping cart total, is first read into the program from some unspecified external source. Perhaps the user has entered some data from which the order amount is computed by another algorithm or perhaps the order amount is read from a database. After reading the order amount, the shipping cost variable is then bound to the value 0. This can be understood as a default value that may be changed to $10 if the order amount is less than the $40 amount required by the business policy of the online store. Of course, the computer needs to make a decision to either bind shipping cost to $10 or not, a decision that is denoted by the diamonds of the flowchart. The topmost diamond splits into two separate paths such that the computer will follow exactly one of the paths. The path labeled as "orderAmount < 40" will be followed if the logical condition is satisfied. This condition corresponds to the yes/no question "Is the order amount less than 40?" If the answer to this question is yes, expressed as a logical value of true by the computer, then the shipping cost takes on the value 10. If the logical condition is not satisfied, then the path labeled as "orderAmount >= 40" will be taken.

Although a flowchart is useful to visualize the order in which actions will take place in an algorithm, most algorithms are written in a textual programming language. Flowcharts are graphical in nature and can be cumbersome to draw and arrange such that the shapes, lines, and arrows are easily followed. Programming languages, by contrast, consist of text that is entered in a linear fashion from beginning to end. Programming languages also tend to be highly structured, meaning that phrases and words must be written in strictly followed patterns. In this textbook we will describe a small set of simple textual patterns that correspond to the elements of a flowchart. Although the language we describe is not a real programming language, it is very similar to many real programming languages.

The syntactic pattern for expressing a one-way selection is shown in Figure 5.8. The highlighted and capitalized phrases are simply placeholders that must contain meaningful phrases when the phrase is used in an actual algorithm. In particular, the CONDITION must be a formula that produces a logical value of either true or false. The ACTIONS must be a sequence of valid actions. The meaning of the one-way selection is that the sequence of ACTIONS is executed whenever the CONDITION is true; otherwise the ACTIONS are not executed.

One-Way Selection

```
if CONDITION then
    ACTIONS
endif
```

FIGURE 5.8   Syntactic pattern for writing a one-way selection.

Using this notation, we can translate the flowchart of Figure 5.7 into a textual program. The variable orderAmount is associated with a value that is generated by a shoppers input. Since the value is not known when the program is written, only when the program is executed, there must be a formula to generate this amount. We will assume that the expression shoppingCartTotal() produces this value for us. This expression might query a database or might add items together to compute the result. The details of how this expression produces a value should not concern us; rather we only need to understand that the expression produces the total order amount when the expression is evaluated.

After line 1 is executed, the shipping cost is associated with the value 0. The program will then determine whether the condition given as "orderAmount < 40" is true or false. If the condition is true, the shipping cost variable is changed to the value 10. It should be apparent how the graphical flowchart corresponds precisely to the written form of this algorithm.

5.3.2.2 Two-Way Selection

Although the algorithm of Figure 5.9 correctly computes the shipping cost for any order amount, you might notice that the algorithm is somewhat inefficient since it will sometimes require two bindings when only one binding is ever necessary. If, for example, the order amount is less than 40 then the shippingCost is bound to 0 and then almost immediately the binding is changed to the value 10. We can use a two-way selection to better and more efficiently compute the shipping cost. In a two-way selection, the computer must choose exactly one of two paths to follow. This is a natural way to express the computation of shipping cost since there are exactly two possible shipping costs and exactly one of them must be chosen.

A two-way selection for computing the shipping cost is shown in Figure 5.10. It is visually apparent from the flowchart that exactly one of the two alternatives will be executed. There will never be a situation where

Program to Compute Shipping Cost

```
1.  orderAmount ← shoppingCartTotal()
2.  shippingCost ← 0
3.  if orderAmount < 40 then
4.      shippingCost ← 10
5.  endif
```

FIGURE 5.9 Program for computing the shipping cost given an order amount.

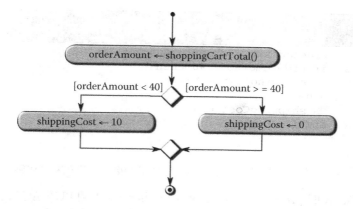

FIGURE 5.10   An improved flowchart for computing the shipping cost.

no action is executed. There will never be a situation where both actions are executed. In this algorithm, the order amount is read from an external source and then a decision is made to either bind the shipping cost to 0 or bind the shipping cost to 10. Comparison with the previous flowchart shows that the action binding the shippingCost to zero has been moved to be part of the selection statement itself.

We will describe a two-way selection statement using the textual phrase shown in Figure 5.11. Just as before, the highlighted and capitalized elements are merely placeholders into which real phrases must be placed. More specifically, the CONDITION must be a formula that produces a logical value while both the IF-TRUE-ACTIONS and IF-FALSE-ACTIONS must be a sequence of valid actions.

When a two-way selection statement is executed, the meaning is to execute either the IF-TRUE-ACTIONS or the IF-FALSE-ACTIONS. Which of these two actions we select depends on the CONDITION. If the

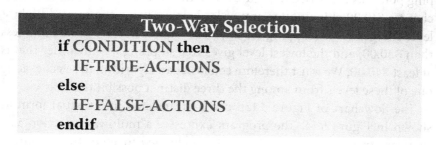

FIGURE 5.11   Syntactic pattern for a two-way selection.

Program to Compute Shipping Cost

```
1.  orderAmount ← shoppingCartTotal()
2.  if orderAmount < 40 then
3.     shippingCost ← 10
4.  else
5.     shippingCost ← 0
6.  endif
```

FIGURE 5.12   Program for computing the shipping cost given an order amount.

CONDITION produces a value of true, then the IF-TRUE-ACTIONS are executed, otherwise the IF-FALSE-ACTIONS are executed.

Using this notation, we can now write a program that more efficiently expresses the shipping cost policy for our online shopping site. Figure 5.12 shows how the flowchart of Figure 5.10 can be expressed as a computable algorithm.

After the orderAmount is obtained in line 1, the program will select either the action on line 3 or the action on line 5 depending upon the condition. In this example, the condition is given as "orderAmount < 40". When this condition is true the computer executes the name binding of line 3; a binding that associates the value 10 with the shipping cost. If the condition produces a value of false, the name binding of line 5 is executed; a binding that associates the value 0 with the shipping cost. The logic and flow of statements in the computable algorithm of Figure 5.12 are identical with the flowchart of Figure 5.11.

5.3.2.3 Multiway Selection

We often need more flexibility than selecting one alternative between only two possible actions. Perhaps the online shop has a three-level shipping policy as expressed in Figure 5.13a. In this example, the highest level charges $10.00 shipping for any order below $20.00, the second highest level charges $5.00 shipping for any order that is at least $20.00 and less than $40.00, and the lowest level gives free shipping for any order that is at least $40.00. We must therefore construct an algorithm to select exactly one of these levels from among the three distinct possibilities.

The flowchart of Figure 5.13b can be translated into textual form as shown in Figure 5.14. The program expresses a multiway selection as a sequence of two-way selections that are chained together by the phrase *elseif*. Notice also that the controlling condition of line 2 is more complex

Shipping Cost Policy	
Order Amount	**Shipping Cost**
$0.00 to $19.99	$10.00
$20.00 to $39.99	$5.00
$40.00 and up	$0.00

(a)

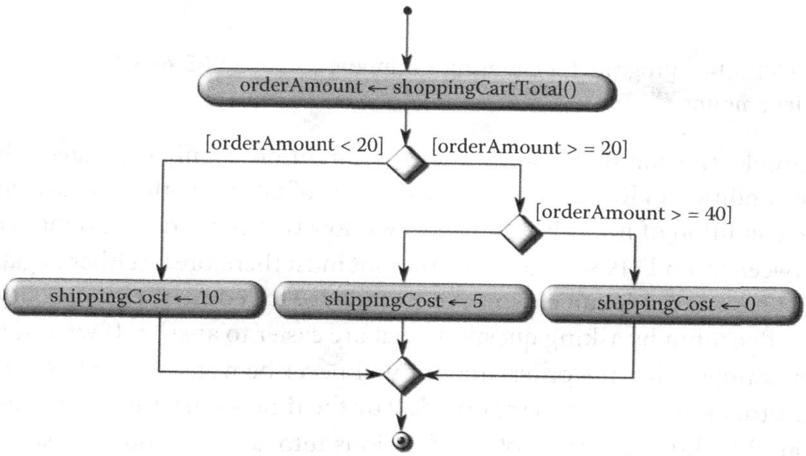

(b)

FIGURE 5.13   (a) Shipping cost policy. (b) Flowchart modeling a multiway selection; selecting one of three alternatives.

since we explicitly check to see if the order amount is at least zero and less than 20 prior to selecting 10 as the shipping cost. This expression corresponds exactly to the way the policy is listed in Figure 5.13b. In similar fashion, the controlling condition of line 4 says that the order amount is at least 20 but less than 40 for the computer to select 5 as the shipping cost. Notice, however, that the final choice is not explicitly controlled by a conditional. The selection statement is formed in such a manner as to select a shipping cost of zero only if no other conditions are applicable. The final action is always selected if neither of the first two actions is taken. In other words, the computer makes a choice that is dependent upon the order in which the conditions are written.

The program of Figure 5.14 can be simplified by understanding that the computer evaluates the conditions from top to bottom. Consider, for

Program to Compute Shipping Cost

```
1.  orderAmount ← shoppingCartTotal()
2.  if orderAmount ≥ 0 and orderAmount < 20 then
3.      shippingCost ← 10
4.  elseif orderAmount ≥ 20 and orderAmount < 40 then
5.      shippingCost ← 5
6.  else
7.      shippingCost ← 0
8.  endif
```

FIGURE 5.14   Program for choosing a shipping cost of $0, $5, or $10 based on the order amount.

example, the condition on line 4. This condition is only evaluated when the condition of line 2 has produced a value of false. In other words, when the condition of line 4 is evaluated we know that the order amount is not between $0 and $19.99. The order amount must therefore be either negative or at least $20. We can exploit this knowledge to reduce the complexity of our algorithm by asking questions that are easier to answer. If we assume, for example, that the order amount will never be negative, we know that the order amount must be at least $20 by the time we ask the second question. By taking the order of the decisions into account and by assuming that the order amount is not allowed to be negative, we can simplify our algorithm as shown in Figure 5.15.

Since we assume that the order amount will never be negative, we eliminate that portion of the criterion from line 2 of Figure 5.15. This makes our algorithm more computationally efficient since the computer does

Program to Compute Shipping Cost

```
1.  orderAmount ← shoppingCartTotal()
2.  if orderAmount < 20 then
3.      shippingCost ← 10
4.  elseif orderAmount < 40 then
5.      shippingCost ← 5
6.  else
7.      shippingCost ← 0
8.  endif
```

FIGURE 5.15   More computationally efficient program for choosing a shipping cost of $0, $5, or $10 based on the order amount.

not need to check twice whether the amount is greater than or equal to zero. In addition, notice that the criteria "orderAmount >= 20" has been removed from the condition of line 4. This is possible since the computer first checks the condition of line 2 and only if that condition does not apply is the condition of line 4 checked. Logically, then, the sequencing of the conditions implies that the order amount must be greater than or equal to 20 whenever the computer checks the condition of line 4.

5.3.3 Repetition

"Wash, rinse, repeat" is a phrase that originated on the back of many brands of shampoo. This phrase has become part of pop culture and is often used as a metaphor for people who robotically follow instructions without critically thinking about what they are doing. Of course the point of this sarcastic metaphor is that anyone following these instructions will literally spend the rest of their lives endlessly repeating the same steps over and over; at least until they run out of shampoo.

Nonetheless, in everyday life we must often repeat a sequence of actions in order to achieve some greater outcome. You paint a wall, for example, by repeatedly dipping your brush into the paint can and applying the paint to the wall. You construct a fence repeatedly digging a hole and inserting a fence post. You even sing a song by repeatedly singing a verse followed by the chorus. Just as many real-life activities require *repetition*, computers often must also execute a sequence of actions repeatedly in order to achieve some greater outcome. In fact, one of the great strengths of a computer is its ability to robotically and quickly repeat actions without complaint!

A *loop* is a control structure that repeatedly executes a sequence of actions. A *while loop* is a type of loop where a sequence of actions is repeated as long as some logical condition holds. All while loops will have the flowchart structure shown in Figure 5.16. When the while loop is first encountered, we must determine whether to perform some action or

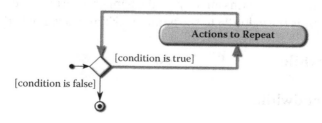

FIGURE 5.16   Flowchart structure for a while loop.

FIGURE 5.17   Flowchart that uses a while loop to model the process of grilling a steak.

sequence of actions at least one (more) time. This decision is made by asking a yes/no question. If the question is answered affirmatively, then the actions are performed after which we again ask whether we must repeat the sequence at least one more time. The repeating loop is highlighted by the red arrow.

Using everyday English we might describe a repeated sequence of actions with a phrase similar to "while the steak has not yet reached a temperature of 135°F, cook for three minutes." In this example, the logical condition is "the steak has not yet reached 135 °F" and the repeated action is "cook for three minutes." A flowchart that models this process is shown in Figure 5.17.

A while loop will be expressed using the notation shown in Figure 5.18 where the CONDITION and the ACTIONS contain elements that make the loop meaningful. In particular, the CONDITION must be an expression that produces a logical value and the ACTIONS is a meaningful sequence of actions. The sequence of actions of a loop is often referred to as the *loop body*. The condition controls how many times the loop body actions will be executed.

Figure 5.19 uses this notation to express a computer program that models the process of grilling a steak. In this program we assume that the steak is placed on the grill at a room temperature of 75° and that for every three minutes the steak remains on the grill the temperature increases by 13°. The first line binds the value 75 to the name steakTemp; a variable that models

while CONDITION do
    ACTIONS
endwhile

FIGURE 5.18   Syntactic pattern for a while loop.

Grill a Steak Program

```
1.  steakTemp ← 75
2.  while steakTemp < 135 do
3.      steakTemp ← steakTemp + 13
4.  endwhile
```

FIGURE 5.19 An algorithm for modeling the temperature of a steak as it is grilled.

the current temperature of the steak. Of course, the current temperature of the steak will increase with the amount of time that the steak remains on the grill. The loop is controlled by the logical condition "steakTemp < 135" and the repeated action is expressed as "steakTemp ← steakTemp + 13". In other words, the loop repeatedly increases the steak temperature in increments of 13 degrees until the minimum desired temperature is attained.

We can gain further insight into the meaning of a while loop by listing the sequence of individual actions that occur when the algorithm is actually executed by a computer. Figure 5.20 gives a list of each action the computer takes when executing the program of Figure 5.19 and also shows the computational state that results from each individual action. The steak temperature is initialized to the value 75. After this initial action, the program repeatedly decides whether to leave the steak on the grill for three more minutes (the result being to increase the steak temperature by precisely 13 degrees). Since, at this point in time, the steak temperature is less than 135, the program decides to grill for three more minutes; and thus the steak temperature changes from 75 to 88. After several times through this process, the steak temperature finally increases to a temperature that is not less than 135. In that case, the program decides that the steak should not be grilled for another three minutes and the loop terminates.

The algorithm for grilling a steak computes the final temperature of the steak at the conclusion of grilling. The final temperature may not be exactly 135 but, as the example illustrates, may be higher than 135 depending upon the starting temperature and the amount by which the temperature rises within a three-minute span.

Line number	Comment	State after execution
1	Bind 75 to steakTemp	{steakTemp=75}
2	The condition steakTemp < 135 is true (therefore repeat)	{steakTemp=75}
3	steamTemp = steakTemp + 13	{steakTemp=88}
2	The condition steakTemp < 135 is true (therefore repeat)	{steakTemp=88}
3	steamTemp = steakTemp + 13	{steakTemp=101}
2	The condition steakTemp < 135 is true (therefore repeat)	{steakTemp=101}
3	steamTemp = steakTemp + 13	{steakTemp=114}
2	The condition steakTemp < 135 is true (therefore repeat)	{steakTemp=114}
3	steamTemp = steakTemp + 13	{steakTemp=127}
2	The condition steakTemp < 135 is true (therefore repeat)	{steakTemp=127}
3	steamTemp = steakTemp + 13	{steakTemp=140}
2	The condition steakTemp < 135 is **false** (therefore stop)	{steakTemp=140}
	Since the condition is false we are done with the program. The variable steakTemp is bound to the value 140.	{steakTemp=140}

FIGURE 5.20   The computational state of the steak grilling program.

A small addition to the algorithm will allow us to compute additional information that a grill chef might require. Perhaps the chef is cooking a large meal that includes steak, baked potatoes, and steamed broccoli. The chef needs to know the time required for grilling the steak in order to ensure that all three dishes are completed at about the same time.

To compute the length of time required, we can introduce a new variable that represents the number of minutes of grilling that have elapsed at the conclusion of every three-minute span of time. This variable should obviously be initialized to zero prior to entering the loop since, at that point in the process, the steak has not yet been put on the grill. Each execution of the loop should increase the variable by three minutes since our model states that the temperature is checked in three minute intervals. Once the loop terminates, the value of the variable will then contain the length of time required to grill the steak. This modification is shown in Figure 5.21 where the variable minOnGrill is initialized to zero in line 2 and incremented by three in line 5.

Computer programmers will quickly recognize that the loop of Figure 5.21 is effectively a counting loop. A *counting loop* is one that contains a variable

Time to Grill a Steak Program

```
1.  steakTemp ← 75
2.  minOnGrill ← 0
3.  while steakTemp < 135 do
4.      steakTemp ← steakTemp + 13
5.      minOnGrill ← minOnGrill + 3
6.  endwhile
```

FIGURE 5.21   Program to compute how long it will take to grill a steak.

to keep track of the number of times the loop is actually executed. In our example, the variable minOnGrill effectively counts the number of times that the loop executes, although the count is maintained as a multiple of three. Prior to executing the loop body, the variable is initialized to zero, thus denoting that the loop has been executed zero times. After the loop body is executed for the first time, the minOnGrill variable holds the value 3, denoting that the loop has been executed 1 time (1 multiple of 3). After the loop is executed for the second time, the minOnGrill variable holds the value 6, denoting that the loop has been executed 2 times (2 multiples of 3).

The algorithm of Figure 5.21 also exhibits a design pattern that all computer programmers are familiar with. This design pattern states that all well-written loops must have three well-defined elements.

1. Initialization—Every variable that occurs in the loop must hold the correct value prior to entering the loop. The initialization of the loop in Figure 5.21 is performed on lines 1 and 2 where we state that the steakTemp immediately prior to the grilling process is 75 degrees and that the steak has been on the grill for 0 minutes.

2. Condition—The condition that determines when to repeat the loop must be precise. Since the chef requires the steak to reach 135 degrees, the condition, given on line 3, indicates that the process will continue as long as steakTemp < 135.

3. Progress—The actions that are repeatedly executed must in some way make progress that allows the loop to terminate. Since the condition of Figure 5.21 implies that the loop will terminate only when

the steakTemp is not less than 135, the loop makes progress toward termination by increasing the steakTemp by 13 with every repetition. This repeated action ensures that the loop will eventually terminate since the steakTemp must eventually exceed 135.

5.3.3.1 Infinite Loops

Consider making one small change in the program of Figure 5.21. Since we know that the steak is done when it reaches a temperature of 135 degrees we decide to reformulate the terminating condition to ensure that we continue to cook the steak as long as the temperature of the steak is not equal to 135 degrees. This change is reflected on line 3 of Figure 5.22 where we change the less than ($<$) to a not equal ($\neq$).

The algorithm of Figure 5.22 is logically flawed since the loop does not make progress toward termination. Notice that the loop will only terminate if the steak temperature is precisely 135. Our program, however, is written in such a way that the steak temperature will never have a value of precisely 135. The steak temperature will instead take on the values of 75, 88, 101, 113, 126, 139, 152, 175, and so forth. The steak temperature will therefore increase without limit; becoming computationally hotter than any object in the universe! In addition, the time required for grilling the steak will also grow without bound.

Figure 5.22 is an example of an *infinite loop*. An infinite loop is a loop that will never terminate because the loop never makes progress toward termination. In other words, when we consult the condition of the loop to determine whether the loop should repeat, the answer is always "yes ... let's do it again!" As we mentioned earlier in this chapter, "wash,

Time to Grill a Steak Program (incorrect)

```
1.  steakTemp ← 75
2.  minOnGrill ← 0
3.  while steakTemp ≠ 135 do
4.      steakTemp ← steakTemp + 13
5.      minOnGrill ← minOnGrill + 3
6.  endwhile
```

FIGURE 5.22 An incorrect program to compute how long it will take to grill a steak.

```
module NAME() is
    ACTIONS
endmodule
```

FIGURE 5.23 The syntactic pattern for naming a subprocess as a module.

rinse, repeat" is a phrase found on the back of many brands of shampoo. This phrase is another example of an infinite loop, at least from a computational point of view, since there is no condition that ever allows us to stop the repetition. An infinite loop is almost always* a logical error since it violates the requirement that a program halt in a finite number of steps. Infinite loops are the source of many mistakes in real-world software.

5.3.4 Modularization

We began this chapter by asserting that any executable algorithm must be expressed as a sequence of possible, or computable, actions. We have described several computable actions that include binding, selection, and repetition along with ability to perform basic arithmetic. We have also described a process for modeling how a steak is grilled; a process that is computable since every discrete action of the process is computable. *Modularization* is a vital element of programming that allows us to define new computable actions by assigning a name to some computable process.

Algorithms can be modularized by breaking them into independent subprocesses. We can, for example, assign the name "grillSteak" to the program of Figure 5.21 and we can then invoke the entire steak grilling algorithm by simply referring to its name. A *module* is a named subprocess. Figure 5.23 gives the syntactic pattern for naming a subprocess as a module.

Figure 5.24 shows how the steak grilling program can be turned into a module by naming the program. In this case, the name grillSteak is associated with the process of grilling a steak. Once a module is defined, we can execute the subprocess by writing the name of the process and appending a pair of empty parenthesis. The notation "grillSteak()", for example, is understood to mean "execute the process named grillSteak."

---

* In limited situations, infinite loops can be logically correct and useful. These situations are beyond the scope of our text.

Grill a Steak Module

```
1.   module grillSteak() is
2.       steakTemp ← 75
3.       while steakTemp < 135 do
4.           steakTemp ← steakTemp + 13
5.       endwhile
6.   endmodule
```

FIGURE 5.24 The grillSteak module.

Modules are vital when writing large programs that consist of many individual parts. Consider, for example, a grill chef who is making dinner for a small private party. We might describe the entire process of making dinner as the execution of a series of modules that involve baking a cake, fixing a salad, making lemonade, and grilling a steak. Each of these modules can be understood as an individual, programmer-defined computable action, and these actions can be invoked by simply referring to them by name. Since these modules have been elevated to the status of a computable action, we can also incorporate them into our flowchart notation. Figure 5.25 shows how a module named makeDinner might be described as a sequence of submodules. Figure 5.25 also shows a flowchart that describes the process.

Well-designed modules should meet several criteria. These criteria generally ensure that the module can be used in a wide variety of larger processes, that the module is self-contained, that any errors caused by a

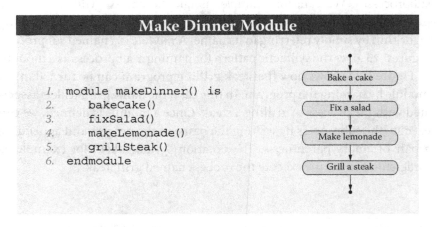

Make Dinner Module

```
1.   module makeDinner() is
2.       bakeCake()
3.       fixSalad()
4.       makeLemonade()
5.       grillSteak()
6.   endmodule
```

Bake a cake

Fix a salad

Make lemonade

Grill a steak

FIGURE 5.25 Making a dinner expressed as a sequence of subprocesses.

module have a limited effect on any system that makes use of the module, and that the module can be used in a wide variety of contexts. These criteria are given as

1. *Understandability*—Every module is self-contained, which implies that it can be fully understood without any knowledge of actions that take place outside of the module itself.

2. *Encapsulation*—Every module affects only the data that it contains. Any errors that arise from a module are also contained within the module.

3. *Composition*—Every module can be incorporated into a larger module without special treatment.

Hosting a private party involves more than just making dinner. The host must, for example, send out invitations, gather up the responses, purchase food, clean the house, and choose background music. We can describe the process of hosting a private party as a module that is composed of these other submodules (Figure 5.26).

Modularity is an extremely powerful technique for designing complex algorithms. Modules allow us to isolate smaller parts of a large process such that those smaller parts are much easier to understand and manage. In addition, since these smaller parts are self-contained, they can be easily inserted into other complex algorithms without needing to customize them or rewrite them. Also, although we must understand precisely what a submodule does, we do not need to fully understand the details of how a submodule actually works.

Host a Party Module

```
1.  module hostParty() is
2.      sendInvites()
3.      collectResponses()
4.      purchaseFood()
5.      cleanHouse()
6.      chooseMusic()
7.      makeDinner()
8.  endmodule
```

FIGURE 5.26   A module for hosting a private dinner party.

5.3.4.1 Module Flexibility

Modules should typically be flexible enough to be used in a variety of conditions. Consider, for example, the steak grilling module of Figure 5.24. This module can only be accurately used if the ambient room temperature is 75 degrees. But what if we are grilling the steak in our backyard and we have allowed the steak to thaw out to the ambient air temperature of 94 degrees? Rather than writing a completely different module, we should rewrite our grillSteak module to make it flexible enough to accommodate any starting steak temperature.

Modules are made flexible by allowing users to feed input values into the code. The module can then respond differently depending on the input values provided by the user. The number of inputs that a module accepts must be defined when a module is written. These inputs are known as formal parameters. A *formal parameter* is a variable where the initial value is bound to the values provided as input to the module. The modules behavior then depends upon the initial value that is provided by the user. The initial values provided by the module user are known as *arguments* or *actual parameter*.

Figure 5.27 gives an expanded syntactic pattern for defining a module. The formal parameters are listed within the parentheses associated with the name. A formal parameter is simply a variable name and a module may have as many formal parameters as convenient for the module's function. Of course, a module may have no formal parameters, as illustrated by all of the modules we have previously described.

We can make our grillSteak module flexible by making the starting temperature an input that is controlled from the outside of the module. This flexibility is shown in Figure 5.28. Note that the module simply defines what actions the computer should take to grill a steak. Although it may initially seem that we have to know the value of the steakTemp variable in order to write a process for grilling the steak, all we really need to know is that the steak has a starting temperature. Of course, if we are ever asked to

> **module NAME(V1, V2, ..., VN) is**
>   **ACTIONS**
> **endmodule**

FIGURE 5.27 The syntactic pattern for defining formal parameters.

Flexible Grill a Steak Module

```
1.  module grillSteak(steakTemp) is
2.      while steakTemp < 135 do
3.          steakTemp ← steakTemp + 13
4.      endwhile
5.  endmodule
```

FIGURE 5.28   A more flexible grillSteak module.

grill an actual steak, we then need to obtain the starting steak temperature to actually carry out the process.

The rewritten grillSteak module now has one formal parameter. This implies that if we ever need to use this module to grill a steak that we must supply one actual parameter. The actual parameter becomes the starting steak temperature for the steak grilling subprocess. If, for example, we want to grill a steak from a starting temperature of 94 degrees because we are grilling on a scorching hot summer day from our back porch in Phoenix, Arizona, we would write

grillSteak(94)

Although the grillSteak module is now flexible enough to be used with any starting steak temperature, we can still increase its flexibility by noting that there are two further circumstances that might not always hold. We might sometimes want to grill a rare steak and hence stop grilling when the steak reaches 130 degrees. We might also want to grill a well-done steak and not stop grilling until the steak reaches perhaps 155 degrees. Also, perhaps it is the case that we have a very thick steak on a slow-cooking grill such that the steak's temperature only increases by 2 degrees for every three minutes the steak is on the grill. These observations imply that we can optimize the flexibility of our grillSteak module by adding two additional formal parameters: the target steak temperature and the increase in temperature for every three minutes of grilling time. These changes are shown in Figure 5.29.

Since the grillSteak module now specifies three formal parameters, we are required to provide three actual parameters when we use the module. In other words, before using the module we must know (1) the starting steak temperature, (2) the target steak temperature, and (3) the amount by which the temperature of the steak will increase with every three minutes

Most Flexible Grill a Steak Module

```
1.  module grillSteak(steakTemp, targetTemp, increase-
    Amount) is
2.      while steakTemp < targetTemp do
3.          steakTemp ← steakTemp + increaseAmount
4.      endwhile
5.  endmodule
```

FIGURE 5.29   An optimally flexible grillSteak module.

left on the grill. Each of the following shows how we can use this grill-Steak module to achieve very different outcomes under very different circumstances.

grillSteak(65, 130, 2)

This causes the steak to be grilled from a starting point of 65 degrees until it reaches a temperature of at least 130 degrees where the temperature increases by 2 degrees for every three minutes of grill time.

grillSteak(94, 155, 13)

This causes the steak to be grilled from a starting point of 94 degrees until it reaches a temperature of at least 155 degrees where the temperature increases by 13 degrees for every three minutes of grill time.

grillSteak(94, 135, 5)

This causes the steak to be grilled from a starting point of 94 degrees until it reaches a temperature of at least 135 degrees where the temperature increases by 5 degrees for every three minutes of grill time.

In summary, we have described name binding, selection, loops, and modules as ways of expressing algorithms and computational processes. We have described how the computational state of an algorithm captures the current values of all data as the algorithm is executed. And finally, we have seen how modules allow us to break large problems into smaller units and how these smaller units can be made flexible through the use of formal parameters.

TERMINOLOGY

activity diagram	loop
actual parameter	counting loop
algorithm	infinite loop
argument	while loop
composition	modularization
computable action	module
computable function	name binding
computational state	program
computer software	programming language
control flow	declarative
control flow statement	functional
encapsulation	imperative
executes	selection statement
expression	multiway
flowchart	one-way
formal parameter	two-way
halt	semantics
identifier	unambiguous
	understandability

EXERCISES

1. Consider writing a software system to determine what you will wear to work. If the morning temperature is above 50 degrees you will wear a pair of olive-green shorts otherwise you will wear a pair of red-flannel sweats.

   a. Write a flowchart that describes how you decide what to wear. Use a variable named "temperature." The first action you chart should take is to initialize the temperature by obtaining the value from the National Weather Service (NWS). You may assume that a module named "getTemperatureFromNWS()" is available for your use. Use a variable named "attire." When the flowchart is completed, the value of this variable must be 0 if you choose to wear shorts and 1 if you choose to wear sweats.

b. Convert your flowchart into module named "whatToWear" using the notation described throughout this chapter.

2. As explained in Exercise 1, consider writing a software system to determine what you will wear to work. In this case you decide to wear overalls if the temperature is below 20 degrees. Otherwise, if the temperature is 50 or below you will wear sweats. In all other cases you will wear shorts.

a. Write a flowchart that describes how you decide what to wear. Follow the directions of Exercise 1, but note that the final value of the "attire" variable must be 0 for shorts, 1 for sweats, and 3 for overalls.

b. Convert your flowchart into a module named "whatToWear" using the notation described throughout this chapter.

3. Assume that a "print" module allows you to write text onto a computer screen. If, for example, you wanted to write the text "Computational Thinking" onto the computer screen, you would express this action as *print(Computational Thinking)*.

a. Write a flowchart that describes how to print the values 0, 3, 6, 9, and 12 to the computer screen. You must use a loop.

b. Write a module named "countBy3" that prints the values 0, 3, 6, 9, and 12 to the computer screen.

4. For each of the following flowcharts, list each action that the computer takes and indicate the computational state that follows from each action.

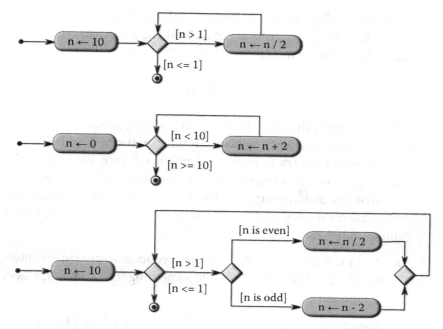

5. Write a module that computes the weekly salary earned by an hourly employee. The module has two formal parameters. The first parameter is named "hoursWorked" and denotes the number of hours that an employee has worked during a one-week pay period. The second parameter is named "hourlyRate" and denotes the number of dollars the employee makes for every hour of nonovertime work. The hourly rate for overtime work, anytime exceeding 40 hours in one week, is 50% above the normal hourly rate. As examples, consider using this module as shown below.

a. weeklySalary(10, 9) should compute the answer 90 since the employee worked 10 nonovertime hours at $9 per hour.

b. weeklySalary(50, 10) should produce 550 since the employee worked 40 nonovertime hours at $10 per hour and 10 overtime hours at $15 per hour.

6. Write a module that computes the volume of a cone. Your module must have two formal parameters. The first parameter is the radius, which is given in inches. The second parameter is the height of the

cylinder and is given in feet. Your module must compute the volume in cubic inches. For a cone of radius R and height H, the volume V is given by the following equation.

$$V = (\Pi * R^2 * H)/3$$

7. Write a module that plays the number guessing game. In this game, the computer chooses a random number between 0 and 100. The player must then try to guess the chosen number. Every time the player guesses, the computer will write "Too low" if the guess is less than the chosen number; "Too High" if the guess is larger than the chosen number; and "You Got It" if the guess is equal to the chosen number.

   a. Assume that the module "random" produces a value in the interval [0, 100]. You can then write something like: computerGuess ← random().

   b. Assume that the module "print" as described in Exercise 3 is provided.

CHAPTER 6

Modeling Solutions

The secret to film is that it is an illusion.

—GEORGE LUCAS

OBJECTIVES

- To be able to interpret activity diagrams involving actions and conditions
- To recognize the three forms (sequences, selections, and repetition) of control that constitute control flow in basic activity diagrams
- To be able to create basic activity diagrams for simple algorithms
- To be able identify states and events from an algorithm
- To be able to interpret state diagrams including do, entry, and exit actions

---

Life is full of *models*. Perhaps in your childhood you made plastic models of cars or boats. Models are also used to show off new clothing designs. Economists create models of financial systems, and sociologists model populations.

A model is nothing more than a replica or representation of some object or system. Sometimes models are used to capture things too small, too fast, or too distant to be understood in other ways. A toy model of the planets of our solar system and a chemist's three-dimensional model of molecular structures are useful because they help us to understand things that are impossible to see with our eyes.

Typically, a model relies on abstraction to emphasize important characteristics and remove unnecessary detail from that which is being modeled. Architects often use physical models of their buildings to highlight the aesthetics of the exterior structure. Movie creators use storyboards to model a film in terms of separate scenes.

One subfield of computer science is devoted to software modeling. Such a computer model is sometimes called a *simulation* because it simulates the actual system. Action games and flight simulators used to train airplane pilots are examples of simulations.

This chapter focuses on two different techniques used by software engineers to model algorithms, namely, *activity diagrams* and *state diagrams*. Both of these techniques utilize abstract pictures that depict selected aspects of algorithms. Both are part of the common diagramming standard notation known as the *Unified Modeling Language (UML)*. Use case diagrams and class diagrams (see Chapter 4) are also a part of the UML standard notations.

You should recall that algorithms consist of two main ingredients:

1. Instructions

2. Data

These ingredients explain the primary difference between activity diagrams and state diagrams: one is focused on instructions (activity) and the other begins with data (and its state).

6.1 ACTIVITY DIAGRAMS

An algorithm is a group of instructions for performing some task. The variety of such tasks is limitless. There are algorithms for driving from one city to another, algorithms for assembling a lawn mower, algorithms for ways to catch the most fish, and algorithms for starting a new company. Computer scientists are most interested in those algorithms that lead to software systems, but the design techniques used to develop software algorithms are generally applicable to many types of algorithms.

We say that an algorithm is correct if it performs the intended task (problem). A *correct algorithm* must use the correct instructions. For example, you can not use fishing instructions to solve a problem such as assembling a lawn mower. However, just because you have identified the right instructions for a task does not mean your algorithm is correct; you must still arrange the instructions in a proper order. Imagine the confusion over directions for driving from one city to another if the order of turns is reversed!

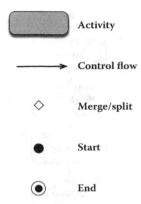

FIGURE 6.1 Symbols used in activity diagrams.

In Chapter 4 the term *control flow* was introduced to mean order of instruction execution, and diagramming control flow is the best way to think of activity diagrams. Some people refer to this kind of diagram as a *flow diagram* or *flowchart*. The picture elements of an activity diagram are quite simple; just five symbols are needed for most diagrams. See Figure 6.1.

A rectangle represents a single activity. Activities correspond to many instructions in a computer program. In nonsoftware algorithms, an activity denotes some action to be performed. The action corresponding to the activity is described inside the activity rectangle. Arrows are used to connect activities and other symbols, indicating the control flow—the order in which the activities must occur. An arrow from the rectangle for Activity A to Activity B specifies that A must be performed prior to B.

Decision/merge diamonds are used in activity diagram locations where the control flow must split or join together as we shall see later. Start and end symbols are used to denote where the algorithm begins or completes execution.

As a first example, consider the activity diagram for a typical morning routine shown in Figure 6.2. This person seems to have a five-step routine to begin the day. After awakening, teeth are brushed, followed by a shower, drying hair, and breakfast, in that order.

Activity diagrams are somewhat abstract in the sense that the instruction descriptions are not always detailed, but activity diagrams are quite precise when it comes to control flow. The morning routine activity diagram includes an order of actions that is unambiguous. Perhaps your

FIGURE 6.2   Activity diagram for a morning routine.

routine is to eat breakfast before showering. Changing the order of activities for such a different routine results in a different algorithm and a different activity diagram is needed. The algorithm depicted in Figure 6.2 clearly indicates that showers precede breakfast.

6.2 SELECTION IN ACTIVITY DIAGRAMS

In Chapter 4 we identified five different forms of control flow:

1. Sequential execution

2. Selection

3. Repetition

4. Control abstraction

5. Concurrency

Activity diagrams can depict all five. In this chapter we explore the first four, while Chapter 11 is devoted to concept of concurrency.

Selection occurs anytime that an algorithm must make a choice. In an activity diagram a selection is depicted as a split in the flow; the control

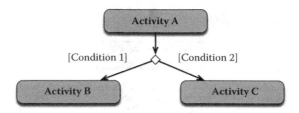

FIGURE 6.3   Using a decision symbol for selection.

flow can follow alternative paths. Figure 6.3 shows how the decision symbol (a small diamond) is used to denote such a split. This diagram shows that following Activity A the algorithm contains a choice of next executing Activity B or Activity C.

Choices in activity diagrams are based upon conditions, where a condition is any statement that must be either true or false (i.e., a logical expression). Each arrow extending from a decision split needs to be labeled with a condition (enclosed in square brackets) for which the associated path is chosen. In Figure 6.3 Activity B is executed in the event that Condition 1 is true following the execution of Activity A. If Condition 2 is true, then Activity C is chosen. A properly written activity diagram crafts the conditions in such a way that either Condition 1 is true or Condition 2 is true, but both cannot be true at once.

Figure 6.4 demonstrates the use of this notation for an algorithm that provides driving directions from University Square Park in Baltimore, Maryland, to Mt. Vernon Square in Washington, D.C. Traffic in this area of the United States can be extremely heavy so this algorithm provides for three potential routes of travel. Regardless of route, the driving directions begin by traveling on W. Baltimore Street.

Presumably, before leaving home the traveler consults a traffic report for the Baltimore-Washington Parkway. In the event that the traffic report is favorable, the driver choses this route (the left path of the diagram) because it is shorter. However, the diagram shows that the arrow to the right should be followed if the traffic report for the parkway is bad. This secondary route follows I-95 and leads to a second decision of whether to exit at Exit 22B or continue to Exit 27. The second selection also depends upon traffic report conditions, comparing highways US-29 and US-50.

Notice that Figure 6.4 also illustrates the use of the diamond-shaped symbol for merging within an activity diagram. In this case the merge occurs in the path from taking Exit 22B off I-95. This exit causes the driver to join the Baltimore-Washington Parkway, the same highway as used in

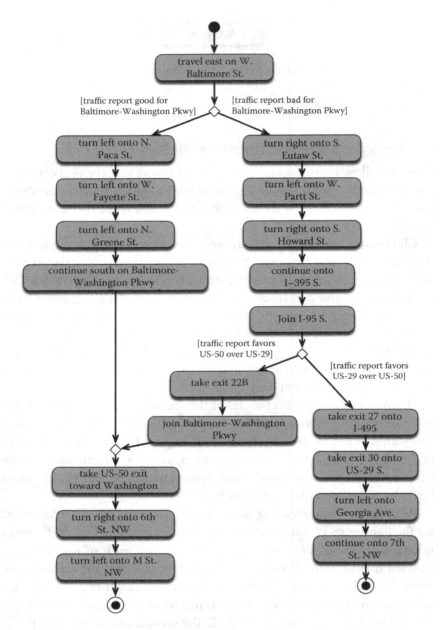

FIGURE 6.4 Activity diagram for directions from Baltimore to Washington, D.C.

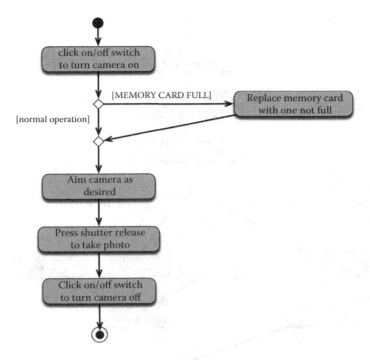

FIGURE 6.5   Activity diagram for taking a photo.

the route of the diagram's leftmost path. Because two routes share the same highway at this point, they also share the same three last activities, beginning with exiting onto Highway US-50. The merge diamond is a way to describe such situations in which two different paths come together for common future activities. Unlike decision diamonds that require condition labels on outgoing arrows, a merge diamond never includes conditions because there is decision to be made in merging.

The decision/merge symbols also work well in depicting optional activities. Figure 6.5 illustrates with operating instructions for taking a photo with a simple point-and-click camera. The optional activity in this case replacing the memory card, because this action is unnecessary unless the card is full.

Not every selection involves just two choices, as shown in the examples so far. A decision diamond may have more outgoing arrows in order to handle more choices. Figure 6.6 contains an activity diagram for the checkout process of a typical online retailer. After an online customer has pressed the Select Payment button, a choice of payment method

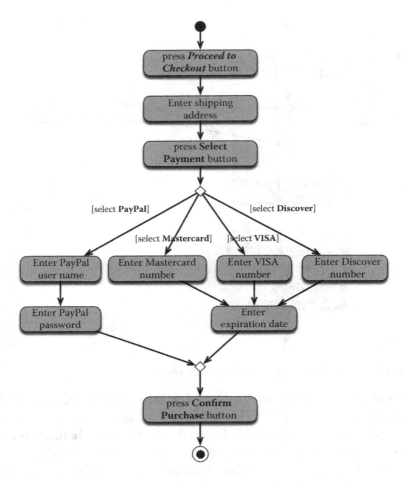

FIGURE 6.6   Activity diagram for checkout process of online retailer.

(PayPal, MasterCard, VISA, or Discover) must be made. This represents a four-way selection.

6.3 REPETITION IN ACTIVITY DIAGRAMS

The decision/merge symbol also serves as a notation for depicting algorithms with repetition. Figure 6.7 demonstrates. This algorithm describes how to place a call to a specific phone number on a typical cell phone. The first activity for placing a call is to turn on the cell, then the user must press the proper button to initiate the telephone app.

The repetition in this algorithm occurs when the caller dials a particular phone number. Notice that the activity diagram pictures this as a

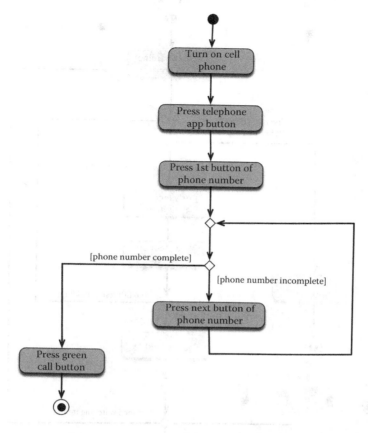

FIGURE 6.7   Activity diagram for placing a cell phone call.

choice based upon whether the phone number is complete or incompletely dialed. In the case that the number is incomplete, the diagram indicates that the caller presses the next button and returns to another choice based upon phone number completeness.

Repetition in an activity diagram is always drawn as an arrow that returns to a location that was encountered earlier in the algorithm. In other words the path of arrows forms a loop, which explains why many computer scientists use the word *loop* to mean algorithmic repetition.

Figure 6.7 shows that arrows sometimes connect decision and merge symbols with no intervening activities. It might be tempting to combine the decision and merge into a single diamond, but it is generally a bit more readable to avoid using dual-purpose diamond symbols.

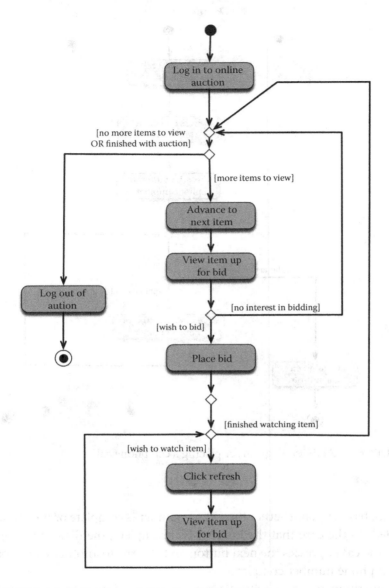

FIGURE 6.8   Activity diagram for an online auction.

A more complex use of repetition is present in the activity diagram of Figure 6.8. The algorithm described by this diagram is for a kind of online auction. This auction only puts a few items up for bid at any point in time. The user is allowed to browse through the items (displayed along with their current highest bid) one item at a time.

The auction activity diagram contains three instances of repetition. One loop describes the user advancing from one item to the next and terminates when there are no more items or when the user is finished viewing auction items. A second loop surrounds the first loop, allowing the user to place a bid before viewing the next item. The bottom loop allows a user to find out whether someone has placed a higher bid by refreshing the current item.

6.4 CONTROL ABSTRACTION IN ACTIVITY DIAGRAMS

Activities do not need to be atomic actions. More complex activities can be abstracted as a rectangle. Zooming in on the rectangle expands it into subactivities. For example, the activity named "log in to online auction" from the Figure 6.8 activity diagram is probably an abstract reference to a more detailed algorithm. Figure 6.9 expands this activity into more detail.

The box in the upper left of the activity diagram in Figure 6.9 shows the subalgorithm represented by the log in activity rectangle. This box contains a five-step sequence that is the subalgorithm. Subalgorithms may contain all of the notations that are permitted in any other activity diagram, including their own starting and ending symbols. The diagram intentionally makes the subalgorithm appear as a blowup of the more abstract activity rectangle. Software applications for manipulating activity diagrams often use zoom-in and zoom-out commands to manage such subalgorithms.

In order to indicate that an activity can be expanded into a subalgorithm, a ⊬ symbol is included in the lower right corner of the abstracted activity in the diagram.

6.5 STATES AND STATE DIAGRAMS

Activity diagrams are useful for diagramming the behavior (control flow) of algorithms. However, modern computer applications often rely on the concept of *state*, suggesting an alternative kind of design diagram.

Physical matter can be in one of three states: solid, liquid, or gaseous. Bits of computer memory can be in one of two states: 0 or 1. A home's climate control system is in one of three states: off or heat or cool.

It is *computational state* that is of the most concern to computer scientists. Imagine that you could take a snapshot of the bit values of every piece of your computer's memory. If so, this snapshot would capture the complete computer's state. Of course, this state is likely to change to a different state almost instantaneously.

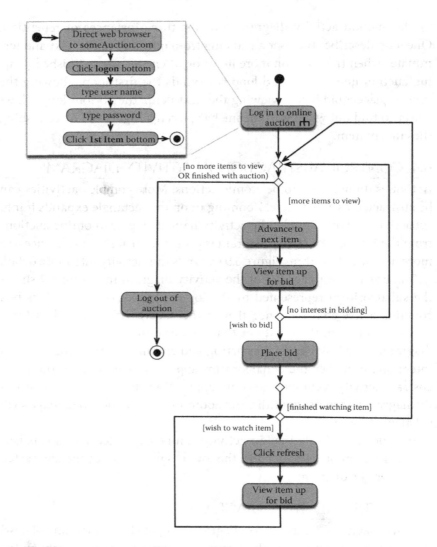

FIGURE 6.9 Activity diagram for an online auction with subalgorithm.

When considering what constitutes state in algorithmic design, it is best to think more broadly about general situations or circumstances from the user's perspective. Software applications are often organized around states in which each state is a separate situation with its own unique conditions.

Applications designed for Internet delivery are especially state dependent. For example, a web page can often be thought of as a separate state that contains its own collection of potential commands, in the form of buttons, links, or tabs. Suppose that an airline application allows users to check flight status, make reservations, and investigate personal information regarding pending flights and frequent flier miles. We can think of these three broad functionalities as three separate states:

- A state for checking flight status
- A state for updating reservations
- A state for checking and changing my account

A user selects some button, tab, or menu entry to enter each of these states, and each state has its own unique web page(s), allowing the user to perform different tasks associated with that state.

State diagrams use many of the same notations as activity diagrams. However, the meanings of the symbols are quite different for activity and state diagrams. See Figure 6.10. Rectangles symbolize states in a state diagram, and arrows denote potential transitions from one state to another. The symbols used at the beginning and ending of a control flow in an activity diagram are also used to indicate a starting state and ending state for state diagrams.

Figure 6.11 shows a simple initial example of a state diagram. This state diagram depicts the three possible states of water: in a liquid form, a solid form (Ice), and a gaseous form (Water Vapor). The four transitions illustrate how this kind of matter changes state. For example, cooling liquid water to temperature below 0°C causes water to freeze into ice. Similarly, heating the liquid form of water above 100°C results in a transformation from liquid to gas. This particular state diagram has the somewhat unusual property of needing no start or end state.

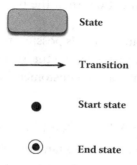

FIGURE 6.10   State diagram notations.

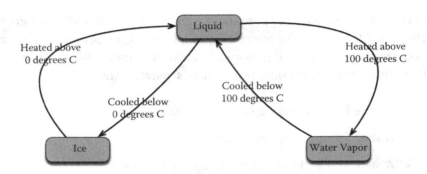

FIGURE 6.11   State diagram for water transitions.

The labels used on transitions describe *events* where an event is a condition or circumstance. The transition labeled by an event applies only when the event occurs. For example, the only time that ice transitions to liquid is when the event "heated above 0 degrees C" occurs.

The state diagram of Figure 6.12 illustrates a more complex system: the behavior of a filling station's gasoline pump. This particular pump requires the use of a credit card. Most of the events described in this diagram would be performed by the filling station customer. Such events include swiping a credit card, inserting the pump nozzle into the gas tank while engaging the filling lever, and disengaging the filling lever while returning the nozzle. The owner of the gas station is also responsible for performing a couple of different events, namely, turning the pump on and off. The pumping system itself is responsible for events involving credit card validation. Yet another way an event occurs is merely due to the passage of time. For example, there is a five-second delay after this pump prints a receipt before the welcome screen allows a customer to swipe the next credit card. Similarly, if the customer fails to engage the nozzle within one minute of credit card validation, then the pump returns to the welcome state.

Figure 6.12 demonstrates that it is possible for a transition to begin and end at the same state. In this example the Warning state has a transition loop to itself in the event that the customer fails to disengage the filling lever properly.

6.6 INCLUDING BEHAVIOR IN STATE DIAGRAMS

State diagrams, like those shown so far, are helpful in understanding the potential changes within a system. However, without more detail these diagrams fail to explain the behavior of the system. It is possible to add

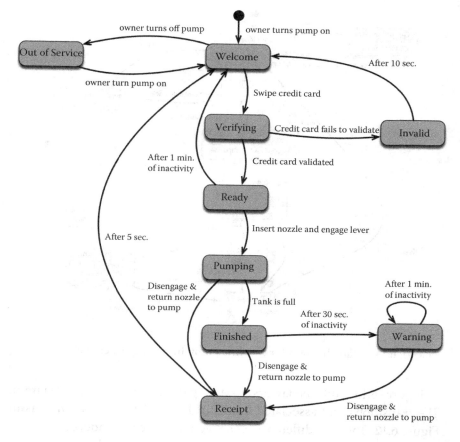

FIGURE 6.12   State diagram for gasoline pump.

system behavior by actions within the states. The correct state diagram notation for such actions divides states into two compartments: the top compartment names the state and the bottom compartment describes actions associated with the state. Each state can include up to three different actions of the following types:

- *entry*—This defines an action that is performed upon each transition into the state.

- *exit*—This defines an action that is performed upon each transition out of the state.

- *do*—This defines an action that is performed continuously while in the state.

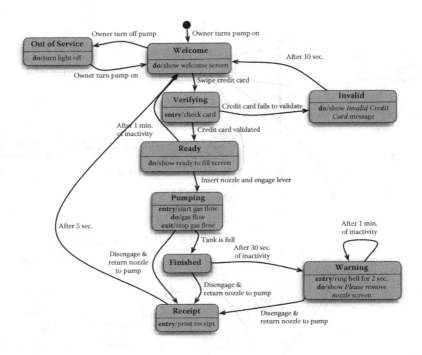

FIGURE 6.13   State diagram for gasoline pump including system behavior.

Figure 6.13 adds behaviors to the gasoline pump state diagram. The transitions and associated events in Figure 6.13 are the same as in Figure 6.12. The only difference is that actions have been added.

An *entry* action is appropriate whenever the action needs to occur upon a transition into the state. Entering the Verifying state causes the customer's credit card to be checked for validity and entering the Receipt state cause the customer's receipt to be printed. *Exit* actions occur whenever a transition is made from a state. Sometimes an equivalent state diagram can be drawn by substituting an exit action of one state with an entry action to a preceding state. For example, the exit action of the Pumping state could be removed by making it into an entry action in both the Finished and Receipt states. Also, note that entry and exit actions are to be performed every time the associated transition is made. Therefore, the alarm bell from the Warning state must ring again for every passing minute of inactivity.

A second example of using entry, exit, and do actions is given in Figure 6.14. This state diagram describes the user authentication system of an ATM. The user of the ATM begins by activating the machine perhaps

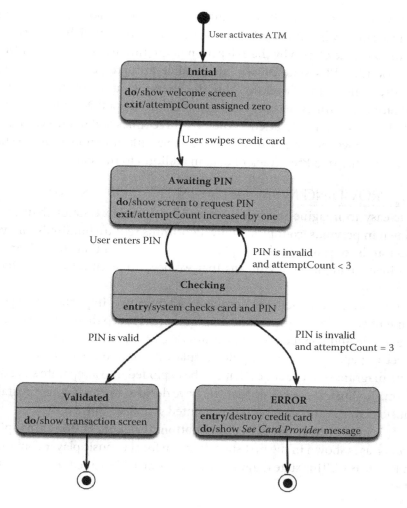

FIGURE 6.14 State diagram for ATM authentication.

by pressing a button on touching the touch screen. While in the Initial state, the ATM is displaying a screen that welcomes the user and asks for a credit card to be swiped. The card swiping is an event causing a transition into the Awaiting PIN state. Immediately before making this transition, the Initial state's exit action is performed. The *do* action of the Awaiting PIN state explains that the user will be requested to enter a PIN at this time. Once the PIN is entered, a transition into the Checking state is made. If the PIN is valid then the next transition is to the Validated state, which serves as the end of this process.

Invalid PINs are handled in a different way. This algorithm makes use of a variable, called *attemptCount*. The attemptCount variable is assigned an initial value of zero by the exit action of the Initial state. Each exit from the Awaiting PIN state causes this variable to increase by one. In other words, while in the Checking state the attemptCount variable will store the number of times the user has attempted to enter a PIN. This particular ATM allows the user to attempt the PIN three times and upon a third failure it destroys the credit card. Using variables, like attemptCount, in state machines helps to keep track of data in addition to the states.

6.7 PROVIDING MORE DETAIL IN STATE DIAGRAMS

It is easy to imagine state diagrams with many more states than those shown in previous examples. Many algorithms require hundreds, or even thousands, of states. Such complexity does not mean that state diagrams are inadequate; it merely means that we need to learn how to use them effectively.

Like other computational problem solving, two important things to remember while designing state diagrams are (1) keep them simple and (2) use divide and conquer. State diagrams are best when they fit on a single sheet of paper, a single computer display. A common way to so restrict state diagram is to use states that can be exploded into a separate state diagram. To illustrate, imagine that you have decided to create a new portable music player in the shape of a five-pointed star. See Figure 6.15.

This music player has five control buttons. Clicking buttons are typical events, as is shown in the first state diagram for the music player contained in Figure 6.16. This state diagram includes a variable, called *songNum*. The

FIGURE 6.15 A star-shaped portable music player.

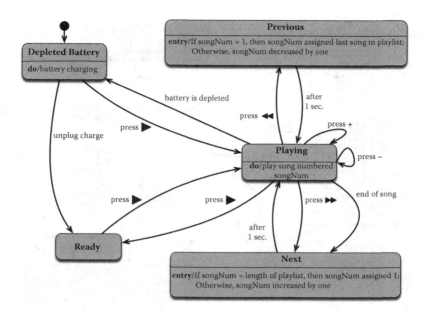

FIGURE 6.16   State diagram for music player.

idea is to use this variable to refer to songs that are numbered from 1 (the first song) to the length of the song list (the last song). Pressing the ►► button during the Playing state causes the songNum, and thereby the player, to advance to the next song, and pressing ◄◄ moves to the preceding song. The state diagram also shows that the song list wraps around in the sense that clicking these buttons to advance off one end of the list causes an advance to the opposite end.

Figure 6.16 also describes that the play (►) button functions both to play and to pause the player. While in the Ready state, pressing ► causes a transition into Playing, and pressing ► in the Playing state causes a transition to the Ready state. Given there is no action in the Ready state, we can presume that no music is playing at this time.

Figure 6.16 describes two aspects of the music player's behavior without much detail. Pressing the + and – buttons produce transitions from the Playing state to the Playing state, but it isn't clear what function this performs. The second vaguely described behavior that is less obvious has to do with charging. It is not quite clear when the user plugged in the charger or what happens if the charger is unplugged so quickly that the battery is still fully depleted.

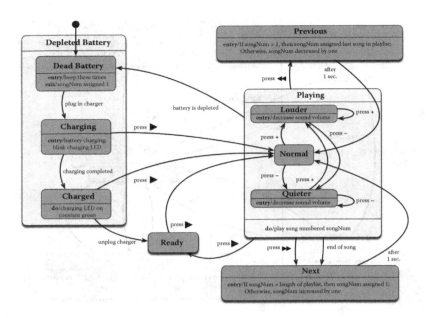

FIGURE 6.17 Exploded state diagram for music player.

The proper way to add more detail to a state diagram is to expand/ explode a state into its own state diagram. Figure 6.17 shows a more detailed state diagram for the music player that explodes both the Playing state and the Depleted Battery state. Notice that these two exploded states are still present, but with state diagrams within. We will refer to the states within an exploded state as the "inner states" and an exploded state as the "outer state."

The Depleted Battery state has been exploded into a separate state diagram consisting of three inner states. These three states detail when the charger should be plugged in and how the music player uses an LED to inform the user about charging. A blinking LED indicates that the device is charging and a constant green LED light indicates that the charge is complete. The Playing state has also been exploded into three states. This explains that the function of the + and – buttons is to raise and lower sound volume.

When states are exploded it is possible for transitions to exit from either an inner or an outer state. When a transition exits from an outer state, the transition applies to all of the associated inner states. For example, if a song ends during either the Louder, Normal, or Quieter state, then a transition to the Next state occurs. Entering transitions must connect to inner states, not outer states.

Figure 6.17 also points out that it is possible for either inner or outer states to include their own entry, exit, and do actions. When an outer state has actions, these must apply to all of its inner states. For example, the do action of the Playing state means that songs continue to play regardless of whether in the Louder, Normal, or Quieter states.

6.8 SUMMARY

Although their focus is somewhat different, both activity diagrams and state diagrams are effective ways to picture and to design algorithms. Both can be applied to any algorithm, although each has its strengths.

Activity diagrams are best when individual states of an application are difficult to discern. Algorithms that seem to exhibit just a single state in which all events are possible at any time tend to follow this pattern. Activity diagrams do a good job of illustrating the order in which instructions are performed. They also illustrate decisions made by the algorithm and repeated activities quite well.

State diagrams tend to be more useful when an application consists of clear and separate states. For example, a software application that uses web pages in which each page has different buttons will create a different set of events for each page. In such a situation the pages suggest states.

State diagrams depict a system as an event-driven algorithm. That is, the system proceeds in response to events, either user events, system-generated events, or events dependent upon the passage of time.

Both activity diagrams and state diagrams support decomposition and abstraction. Sometimes we can zoom in on the actions of an activity diagram to obtain more detail. Similarly, states of a state diagram can sometimes be exploded into their component state diagrams. In both cases it is possible to not zoom or not explode in order to abstract unwanted detail.

Modeling algorithms pictorially is an effective way to communicate the key aspects of the algorithm. Computer scientists and non-computer scientists find activity diagrams and state diagrams to be helpful tools for communicating algorithms of all kinds.

6.9 WHEN WILL I EVER USE THIS STUFF?

It has been said that a picture is worth a thousand words. Not everyone will create computer programs in their lifetime, but we all use algorithms daily and we can all benefit from ways to diagram these algorithms as a way to communicate. The director of a human resource department depicts the hiring procedures and policies in the form of an activity diagram. The art teacher explains the best ways to create ceramic pottery in the form of a state diagram.

Nowhere are these diagrams more important than as a tool for communicating with engineers. Activity and state diagrams serve as an abstract kind of language that allows a design engineer to communicate ideas with customers. If you can interpret an activity diagram, you are equipped to understand what the software engineer is planning for your company's new retail website. If you understand state diagrams, then you can understand the plans of an automotive engineer creating a new car dashboard.

Not only are activity and state diagrams good for designing systems, but they can also be helpful for analyzing systems. If we draw a state diagram for an organization's shipping process, we might discover combinations of states and events that were never considered. An activity diagram of your daily routine might help you plan time for additional volunteer work or a new committee obligation.

TERMINOLOGY

activity diagram	loop
control flow	model
do (state action)	simulation
entry (state action)	state (of computation)
exit (state action)	state diagram
event	Unified Modeling Language (UML)
flowchart	

EXERCISES

1. Which of the following is (are) the rough form(s) of the flow diagram depicting a loop (repetition)?

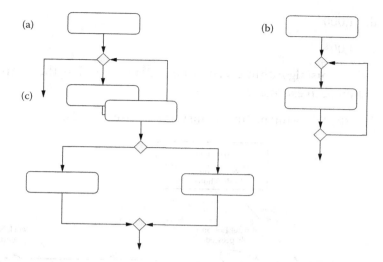

a. Both a) and b) depict loops.

b. All three depict loops.

Use the picture below to complete Exercises 2 through 4.

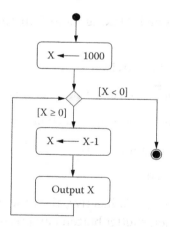

2. Is the picture an activity diagram or a state diagram?

3. What is the first value of X output when this algorithm executes?

    a. 0

    b. 1

    c. 999

    d. 1,000

    e. 1,001

4. When this algorithm executes how many times has the Output X action been executed?

Use the following picture to answer Exercises 5 and 6.

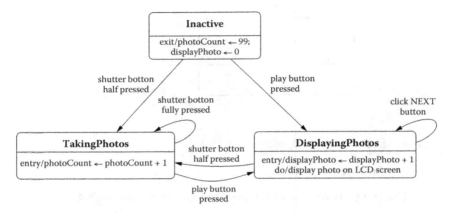

5. The rectangle named "TakingPhotos" in this diagram represents which of these?

    a. A computational activity

    b. One instruction from a computer program

    c. A group of instructions from a computer program

    d. A computer program

    e. A computational state

6. Assuming the camera is inactive, then the user presses buttons in the following order: shutter button half pressed, shutter button fully pressed, play button pressed, shutter button half pressed. What is the resulting value of photoCount?

7. Draw an activity diagram for the following algorithm, assuming that pepperoni and mozzarella are both integer variables.

```
pepperoni ⇐ 10
    mozzarella ⇐ 1
    if  pepperoni ≥ mozzarella  then
    while pepperoni > 8 do
        mozzarella ⇐ mozzarella + pepperoni
        pepperoni ⇐ pepperoni - 1
    endwhile
    else
      pepperoni ⇐ 0
    endif
    mozzarella ⇐ mozzarella + 1000
```

Use the following diagram to answer Exercises 8 through 10.

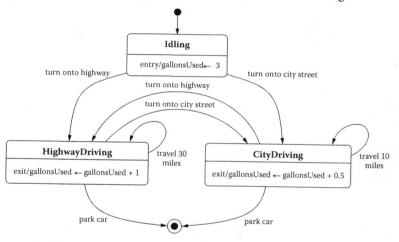

8. What do the rectangles in the diagram represent? (Hint: This is also the name of this kind of diagram.)

9. Using the diagram, how many total gallons of gas are used in a complete trip that begins from idle, followed by 24 miles of city travel, followed by 29 miles of highway travel.

10. Which one of the following is an event in the diagram?

    a. park car

    b. idling

    c. exit

    d. entry

    e. gallonsUsed ← gallonsUsed + 1

CHAPTER 7

Data Organization

In the successful organization, no detail is too small to escape close attention.

—LOU HOLTZ

OBJECTIVES

- To understand the importance of properly naming items
- To understand how a computing system organizes data in memory
- To understand how lists, trees, and graphs can be stored in memory
- To understand how lists, trees, and graphs are correlated to familiar concepts such as family trees, road maps, and organizational charts
- To understand how indexing is used to organize data in memory
- To understand how linking is used to organize data in memory

---

Have you ever had trouble finding an important item in your room because the room was cluttered and disorganized? Have you ever had trouble finding a file on your computer because you could not remember what it was named? Computing systems commonly manage billions of items with every passing moment. A computing system must keep precise account of this data as it flows through the system and this requires close attention to every small detail. In particular, this chapter explains that all data must be well organized and properly identified in order to be usable.

7.1 NAMES

Theodor Geisel wrote and illustrated more than 45 children's books. His books were filled with imaginative scenery and fantastically shaped creatures. Much of his writing used fictional names that created a poetic rhythm to his work and rolled smoothly off the tongue when read aloud. His drawing and watercolors were equally vibrant, carrying an intense and unique style. Theodor Seuss Geisel, also known as Dr. Seuss, once wrote a short story about a woman who had 23 sons each having the name Dave. The following is the opening stanza of this short story, "Too Many Daves"[1].

> Did I ever tell you that Mrs. McCave
> Had twenty-three sons and she named them all Dave?
> Well, she did. And that wasn't a smart thing to do.
> You see, when she wants one and calls out, "Yoo-Hoo!
> Come into the house, Dave!" she doesn't get one.
> All twenty-three Daves of hers come on the run!
> This makes things quite difficult at the McCaves'
> As you can imagine, with so many Daves.

The story goes on to say that Mrs. McCave wished she had named each son differently so that there would not be so much difficulty. The same thing is true of computing. When computational data is inappropriately or confusingly named it becomes very difficult to access or reason about the data. This same principle is not only true in the context of computation but is equally true in real life. Although it can be very difficult to determine the correct name for at item, if we inappropriately name something in the real world, some degree of confusion and difficulty will inevitably follow. If, for example, we mistakenly identify a coral snake (a venomous variety) as a scarlet snake (a nonvenomous variety) we are likely to endure much more difficulty than Mrs. McCave!

There are only a few guidelines that should be followed to properly name items in a computational setting. These guidelines help to ensure that two different people or computing systems can identify, locate, and reason about data without confusion.

1. Names should be unique. A name should refer to only one thing and never more than one. Mrs. McCave, for example, used one name to refer to many sons and this caused confusion at her household. If a name only refers to a single item, then whenever that name is used there is no confusion about what item is being referred to.

2. One item should not have more than one name. If one item has two different names, it can be confusing to communicate with other people (or other computing systems) about that item. If, for example, I refer to the author of "Too Many Daves" as Theodor and you refer to the author as Dr. Seuss; communication may be more difficult than if we both used the same name.

3. A name should be descriptive. When computing with data, the name of an item should describe its role or function within the system. What if we decided to give random names to all of the MP3 music files on our computer? We might, for example, end up with files named *zqiy.mp3*, *gzpu.mp3*, and *gaep.mp3*. Although each name would be unique and no file would have more than one name, the names themselves engender confusion because they are not related to the content of the music. Descriptive names reduce confusion by helping us understand the role or the content of the item. A file named *The Star Spangled Banner.mp3* is preferable to naming that same file *zqiy.mp3*.

4. The name of an item should be related to the location of the item. Computing systems must manage and organize an overwhelming amount of data. The World Wide Web, for example, is a collection of trillions upon trillions of pieces of data and each of these items must not only be uniquely named but also quickly located in order to be useful.

The World Wide Web provides one of the best examples of the importance of proper naming. Even though the web contains trillions upon trillions of pages of information, every piece of this information has a unique name. These names are informally known as web addresses but are formally known as Uniform Resource Locators (URLs). The URL of a web page is technically not the name of the item itself, but is rather a way of locating the item. Nonetheless, in a practical sense, the URL serves as both the name and the location of an item. Consider whether URLs follow the four guidelines for proper naming.

1. Names should be unique. URLs are unique since a single URL will not refer to more than one website. For example, http://www.wikipedia.org/ always refers to the Wikipedia website and not to some weather or news page. If this URL is typed into a web browser, one will always be directed to the main Wikipedia page and not to a variety of randomly determined websites.

2. One item should not have more than one name. Although this guideline can be violated, it rarely ever is. Most websites do not have two different URLs to refer to the same page on the site. It is very rare, for example, to type two different URLs into a web browser and be directed to the same web page.

3. A name should be descriptive. Most URLs contain a description of the information on the web page. Consider, for example, the URL http://en.wikipedia.org/wiki/Coral_snake. This URL is the English language (en) Wikipedia page describing coral snakes. It would be surprising to load this page into a browser and see a list of news items related to recent volcanic eruptions or a description of the top grossing movies of all time.

4. The name of an item should be related to the location of the item. Every URL indicates not only the content of the page but is primarily related to where the page can be located. Every computer on the Internet has a name and the first part of a URL identifies one computer on the Internet. Each computer also has a file system and the remaining parts of a URL correspond to the location of a file in that file system. For example, the URL http://en.wikipedia.org/wiki/Coral_snake indicates, roughly, that the computer named *en.wikipedia.org* has a folder named *wiki* inside of which is a file named *Coral_snake*.

The ability to properly identify and name items is essential when managing large amounts of information. It is also essential in real life since improper names create unnecessary confusion when reasoning about and communicating information. Throughout the remainder of this chapter we will see how large collections of items can be stored within a computer so that the items are easily located. The most common ways of organizing information are as lists, graphs, and trees.

7.2 LISTS

Are you the type of person that makes a list of everything? Perhaps you have a list of your favorite friends, a list of enemies, a shopping list, a list of your favorite movies or favorite foods, a to-do list, a pet-peeve list, a reading list, a wish list, or even a list of all of your lists. Much of the world's data is best organized in list form and deep thought is required to understand how lists can be stored and manipulated by computing systems.

A *list* is a sequence of items that are arranged in a particular order. Consider the following list of the top five most expensive paintings of all time. It is amazing to find out that *The Card Players* was privately purchased for between $250 million to $300 million as of the time of this writing! [2]

1. *The Card Players* by Paul Cézanne

2. *No. 5, 1948* by Jackson Pollock

3. *Woman III* by Willem de Kooning

4. *Portrait of Adele Bloch-Bauer I* by Gustav Klimt

5. *Portrait of Dr. Gachet* by Vincent van Gogh

Any item on the list can be identified by its position in the list. We might, for example, say that *The Card Players* is the number one most expensive painting of all time. We might also say that *Portrait of Dr. Gachet* is the fifth (or number five) most expensive painting of all time. When we use numbers to identify the things in a list we are using a technique known as indexing. *Indexing* associates a unique number with every item in a set of data and therefore allows the items to be identified by their index. Since indices are used to identify data within a list; an index can be understood as a unique name for one of the items in a list.

In this chapter, we will denote an item in a list by using brackets. For example, if Paintings is the name of a list, we denote the ith painting as Paintings[i]. Using this notation, the first painting in the list is denoted as Paintings[1] and the second item in the list is denoted as Paintings[2]. For now we are assuming that the first item in a list has an index of 1, but we will see later that most computing systems assume the first item in a list has an index of 0. For any computing system that uses zero as the starting index, we denote the first item in the list as Paintings[0] and the second item in the list as Paintings[1].

Consider how a list of items can be stored in computer memory. In computing systems the memory is a one-dimensional arrangement of items such that each item is assigned a memory address. All of the data in a computer is stored at some location in memory and each *memory location* is numbered as a list starting from zero. Each memory location can store one *word* of data where a word is the smallest unit of data that is naturally stored by some computing system. Figure 7.1 depicts how each word in memory is stored at a memory address and that these addresses

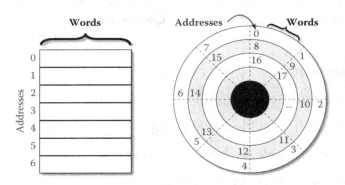

FIGURE 7.1 Memory is linear. Each piece of data in memory is located at a memory address. Memory addresses are numbers or indices. Even CD and DVD memory is organized linearly.

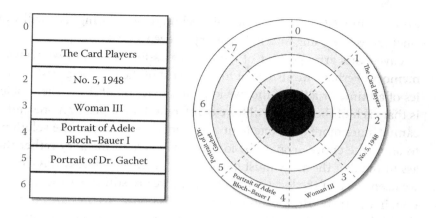

FIGURE 7.2 The list of five paintings can be stored as an array in memory or on a DVD.

are integer numbers. Memory addresses always start at zero and are as large as the amount of memory in a particular computing system.

Although hard disks, compact discs (CDs), and digital video discs (DVDs) are physically shaped as discs, the data stored on these devices is also arranged in a linear manner. Each *disc* is divided into concentric rings known as *tracks*. Each track is divided into arcs known as *sectors*. Each sector is able to store one word of data.* As a disc spins, a single track will scan through a sequence of words whose indices increase linearly.

7.2.1 Arrays

7.2.1.1 Storage

An *array* is perhaps the simplest way of storing a list in memory. An array stores each item at the memory address corresponding to the items position in the list. If we store the list of the top five most expensive paintings, for example, we would store the information as shown in Figure 7.2. Since *The Card Players* is at position one, it is stored at address 1 and since the *Portrait of Dr. Gachet* is at position five in the list, it is stored at address 5. This description assumes that the title of the painting can be stored in one word and that there is only one list that we need to store in memory since storing two lists would require us to store two different things in each

---

* Sectors will actually contain more than one word of data and sectors nearer the center of the disc may contain less data than sectors farther from the center. For our purposes, however, it is sufficient to indicate that the words are arranged in a logically linear manner.

word. Figure 7.2 illustrates that this technique for storing lists will work whether we are using internal memory or even a DVD.

Given this arrangement of data, is it possible to store two arrays in memory? Perhaps we also want to store a list of the five top grossing movies of all time or a list of the five longest movies of all time. The problem is that each of these lists will have a different item at position one and we cannot store more than one item at memory address one. The solution is to allow the entire array to be located anywhere in memory rather than assuming that the array must be located at memory address 1. As long as we keep track of the location of the array, we are still able to access every item in the array. The location of the array is known as the *base address* or the *anchor*, and is the memory address of the first item in the array.

From a computer's perspective, the name of an array corresponds to the anchor of the array. If the name of the array is known, then the anchor is known and all of the array items can be accessed by their index. Figure 7.3 gives an example of two different lists that are stored in memory as arrays. The array named *Paintings* is anchored as location 3 and the array named *Movies* is anchored at location 8. From the viewpoint of the computer, the

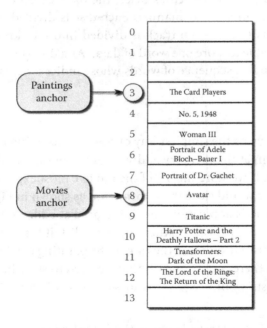

FIGURE 7.3 Two lists stored as arrays in memory. List A is anchored at memory address 3, and list B is anchored at address 8.

name *Paintings* actually means memory location 3 and the name *Movies* actually means memory location 8.

One very important property of an array is that once an array is stored in memory, *the array cannot change its location nor can it change its length.* The array cannot expand to fill more memory nor can it shrink to take up less memory. Figure 7.3 shows one reason why an array cannot expand. In this example, the Paintings array cannot expand since there is another array stored immediately after it.

7.2.1.2 Accessing Array Elements

One of the main advantages of arrays is that we can easily find any item in the array if we know the index, or position, of the item in the list. If for example, we want to find the second item in the list of paintings, we would name that item as Paintings[2]. This name indicates to the computer where the item is located in memory. The computer associates the name Paintings with the anchor (memory address 3) and then takes the number 2 as an offset from the anchor. For this example, the computer moves forward 1 memory location from the anchor to arrive at memory address 4. The ith item in an array A is named A[i] and the memory location of that item is given by the following simple formula.

$$\text{(anchor of array A)} + (i - 1)$$

Every time a computing system looks up an item in an array, the system must perform one subtraction and one addition to find the items address. Since performing these operations slows the computing system, most computing systems adopt the convention that the first item in a list is numbered from 0 rather than 1. This convention is known as *zero indexing*. When using zero indexing, the first item in a list is considered to be at position 0 and the second item in a list is considered to be at position 1 and so forth. If we adopt this convention, we can compute the memory address of the ith item in an array by using the formula:

$$\text{(anchor of array)} + i$$

7.2.1.3 Deleting Array Elements

Lists will almost certainly change over time. Someone will eventually pay more than \$300 million for a painting and some movie will eventually make more money than James Cameron's *Avatar*. If we have a to-do list we will cross things off the list as we complete each task and add other tasks to the list. The two most common ways of changing a list are to delete an item and to insert an item.

Consider how our list of the five most expensive paintings of all time would change if we were to delete *The Card Players* from the list. Since the purchase was a private transaction we might hypothetically discover that the painting had only been purchased for a mere $100 million, which would place it below other more expensive paintings. This would require us to delete *The Card Players* after which we might insert the next most expensive painting at the end of the list.

Deleting an element from an array is very similar to deleting an item from a handwritten list. You would likely first erase all of the items in the list, and then rewrite them once they have been erased. If you also wanted to insert something into the list, you would simply rewrite the entire list in the desired order. This process is obviously not very efficient since we essentially rewrite the entire list any time we delete an item from the list. At least, we rewrite all of the items in the list that follow the deleted item. Figure 7.4 illustrates this process.

Although we have already mentioned that an array cannot expand or shrink, it is possible to both delete items from an array and to insert items into an array. If, for example, we delete the first item in the list of paintings, each of the other elements in the list would have to shift in memory since their position in the list changes as a result of the deletion. The second item in the list must become the first item and the third item in the list must become the second and so forth. The array itself will still consume the same amount of memory locations but the last memory location will be blank after the deletion.

Figure 7.5 shows the various steps required to delete *The Card Players* from the Paintings list of Figure 7.4. We must first move the second item in the list, *No. 5, 1948*, into position 1 of the new list.

FIGURE 7.4 Deleting the first item from a list of famous paintings. The first step is to erase the entire list. The second step is to rewrite the list. Not very efficient.

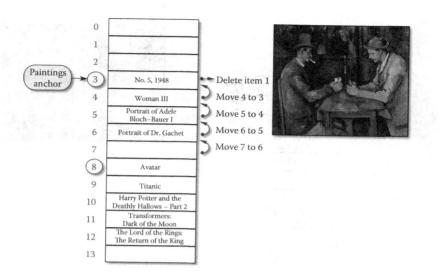

FIGURE 7.5   Deleting the first item in the Paintings list. This requires moving four items in memory starting with earlier items in the list and moving toward the end of the list.

Since the first item in the list must be stored at the anchor, we must move the second item in the list, the item at memory location 4, to the anchor position. Each subsequent item in the list must also be shifted in memory since its position in the list changes as a result of the deletion. Note that the order in which these moves are performed must be from the front of the list to the end of the list, otherwise the operation will not work correctly. For this example, deleting the top item in the list requires four moves.

7.2.1.4 Inserting Array Elements

We can also insert an element into the array, but only if there are blank locations at the end of the array. Consider, for example, what would happen if someone purchased *Bal du moulin de la Galette,* a painting by Pierre-Auguste Renoir, for $350 million. We would then insert this item at the beginning of the paintings list since the purchase price would make the painting the most expensive in history. Insertion at the beginning of the list requires us to first shift every item in the list and then to finally place the entered item into the anchor location. These operations must take place from the end of the list working toward the beginning of the list to prevent data loss. This is illustrated in Figure 7.6 where the first step is

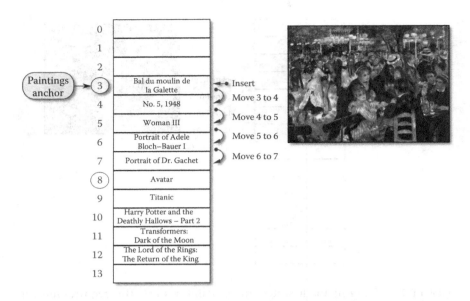

FIGURE 7.6 Inserting an item at the beginning of a list. This requires shifting all items in the list to make room for the newly inserted item.

to move the *Portrait of Dr. Gachet* from location 6 to location 7 after which we move the other elements forward by one memory location. Finally, we insert *Bal du moulin de la Galette* at location 3, thereby making it the first item in the list.

7.2.1.5 Array Summary
Any list can be stored in memory as an array. The main advantage of using an array is that any element in an array can be easily accessed by just knowing the position of the item in the list. The main disadvantages are that the size of the array is fixed and that inserting and deleting items in the list will often require a significant number of steps.

7.2.2 Linking
7.2.2.1 Storage
Linking data in memory is the most common technique for storing a list in memory. When we link data together we form a chain of items that are scattered throughout memory but are nonetheless very well organized and easily manipulated. This type of storage technique is known as a *linked list*.

The fundamental idea behind linking data is very similar to certain forms of geocaching. *Geocaching* is a real-world treasure hunting game where players try to locate hidden treasures, known as geocaches, by using GPS-enabled devices. In a basic geocache game, a person will place a logbook and some special items, the treasure, into a waterproof container. The person will then hide the container, also known as the cache, somewhere interesting and record the GPS coordinates of the cache. These coordinates are then posted on an Internet site and other players attempt to find the cache.

When a player finds the cache, he or she will sign the logbook and perhaps even replace the treasure they find with a treasure of their own. The treasures are not typically valuable objects, but are interesting trinkets or perhaps objects that have personal or sentimental value to the player.

A geocache race, or a multicache, introduces a slight variation into the basic geocaching game. In this variation, each waterproof container will house not only a logbook and treasure but will also contain the GPS coordinates of the next geocache in a sequence of geocaches. Contestants in the race will begin with the GPS coordinates of the first geocache. Only when they find the first geocache will they know what treasure the cache contains and where the next treasure can be found. The winner of the race is the first person to find all items in the race. Players will know they have found the last geocache whenever they find a container without any further GPS coordinates. The race can be summarized as a list of containers where each container holds both a treasure and the GPS location of the next container.

Figure 7.7 illustrates how a geocache race is completed. All participants initially meet at a certain location, which in our example is denoted by the

FIGURE 7.7   Illustration of a geocache race.

yellow dot. Each participant is given the location of the first cache, which in our example is located at (38.6, 97.5). Note that these are the coordinates of the first item in the race, not the coordinates of the starting point. When we reach the first cache, we find a container with a pocket watch and a new set of coordinates, (38.7, 97.9). The pocket watch is the first item in the race and the new coordinates indicate the location of the second item in the race. After reaching the second location we find a container with an ornamental ship anchor and the coordinates of the third cache, (38.9, 97.5). After reaching the third location we find a container with a seashell, but there are no coordinates. Since there are no more coordinates, we realize that we have completed the race. In addition, we now know that the items in the race are (1) a pocket watch, (2) an ornamental ship anchor, and (3) a seashell. We discovered all of this information by starting only with the GPS coordinates of the first item in the race.

The items in a list can be linked together in memory using a technique that is very similar to geocaching. We first need a waterproof container to hold our data. We therefore define a *node*, a (waterproof?) container, to be an adjacent pair of memory cells. The first node value holds an item in the list and the second node value holds the memory location of the next item in the list. By convention, we say that if the second node value holds the number zero, then there are no more items in the list. We define the *anchor* of the list to be memory address of the first node.

Figure 7.8 illustrates how the items in the paintings list can be linked in memory. In part (a) of this figure, we see 16 memory cells and the data that each cell contains. A proper interpretation of this data is shown in part (b) where the cell-pairs that compose a single node are grouped together and the linkages are shown with arcs that connect the nodes. Initially, the only thing that we know is the anchor of the list, the memory location of the first node. In our example, the anchor node at location 3 is composed of the two cells at memory locations 3 and 4.

The first item in the list is stored at location 3 and the location of the next item will be stored at location 4. In other words, location 4 will not contain the next item, but rather it will contain the location of the next item. We note that *The Card Players* is the first item in the list and that the next item in the list is stored at location 12. Moving to location 12 we find that the next item in the list is *No. 5, 1948* and that the location of the next item is 1. Moving to location 1 we find that the next item in the list is *Woman III* and that the location of the next item is 5. Moving to location

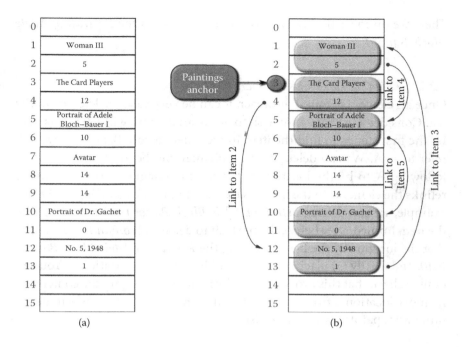

FIGURE 7.8 Storing a list of five paintings by linking the items together in memory. (a) Memory without any interpretive notations, (b) memory with interpretive notations.

5 we find that the *Portrait of Adele Bloch-Bauer I* is the next item in the list and this item is followed by the painting stored at location 10. Moving to location 10 we find that the next item in the list is the *Portrait of Dr. Gachet* and that there are no more items in the list since the location of the next item is 0. Recall that the value 0 is a special value that denotes the end of the list when it is used as a memory location.

7.2.2.2 Accessing Linked List Elements

Perhaps the only real disadvantage to linked lists is the difficulty of accessing an item in the list given only the index of the element in the list. To identify the Nth item in a list requires that we start at the anchor and move through exactly $N - 1$ links. Locating the fourth item in the list of Figure 7.8, for example, requires us to start at the anchor, follow the first link to arrive at location 12, follow the second link to arrive at location 1, and then follow the third and final link to arrive at location 5.

There we discover that the fourth item in the list is the *Portrait of Adele Bloch-Bauer I.*

7.2.2.3 Deleting Linked List Elements

Once an element is found, however, it can be very easily deleted from the list. Consider a list where we wish to delete an item we will refer to as item B. The first step is to find the item immediately before B, an item we will refer to as A. We can delete B by simply making the location of the item following A to be the location of the item following B. This essentially relinks the elements in the list to bypass item B. In the list of Figure 7.8, for example, we can delete *Portrait of Adele Bloch-Bauer I* by (1) finding that the item immediately before the portrait to delete is *Woman III*, (2) noting that the location of the item following the *Portrait of Adele Bloch-Bauer I is 10*, and finally (3) placing the value 10 into memory location 2. You can double check that this process works by mentally changing the contents of memory location 2 from a 5 to a 10 and then following the links to determine what paintings are in the list.

7.2.2.4 Inserting Linked List Elements

Inserting into a linked list is similar to deletion. Consider a list that contains two adjacent items A and C. If we insert item B between items A and C we must simply change the location of the item following A to be the location of B and also make the location of the item following B to be the location of C. A change to these two memory locations is sufficient for insertion.

Consider inserting a new painting into the list of Figure 7.9. Perhaps a rising new artist has recently sold a painting named *Computational Abstraction* that makes it the second most expensive painting in history. We therefore seek to insert this painting as the second item in the list. The first step requires us to identify two unused memory cells that we will use as a list node. Since memory cells 14 and 15 are unused we define this pair of cells as a node.

Since the first value of a node holds the value of the list item, we can immediately place the title *Computational Abstraction* into memory location 14. We must then relink certain items in the list to pass through this newly created node in the correct order. This relinking is accomplished by first placing a 14 into cell 4 since cell 4 contains the location of the node that immediately follows *The Card Players*. Finally, we place the value 12 into cell

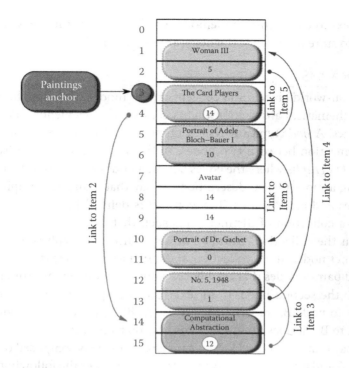

FIGURE 7.9 Inserting an element into a list involves creating a new node (as shown in cells 14 and 15) and changing an existing link to refer to the new node (as shown in cell 4).

15 since *No. 5, 1948* follows the newly inserted item. This process is illustrated in Figure 7.9 where the memory cells that have been changed are highlighted.

7.2.2.5 Linked List Summary

Insertion into a list and deletion from a list are slowed only because we first have to chain to the position where we will either insert or remove. Inserting at the beginning of the list and removing the first element in a list are therefore very fast operations since the first item is the easiest to locate. Thoughtful readers will perhaps question whether an item can actually be inserted at the beginning of a list or even whether the first element in a list can be deleted. Our presentation here does obscure a rather minor detail present in most real programming situations. In real languages, the anchor is allowed to change, which makes operations on the first element very efficient and convenient. In addition, the items in the list do not have to be next to each other in memory, which means that it is easier to find places to store items since the items are smaller in size.

7.3 GRAPHS

Many real-world objects and concepts are be modeled as a graph. A graph is a mathematical abstraction that is composed of elements called nodes and edges. A *node* typically represents some real-world item, and an *edge* is a connection between two nodes. In this text, we limit our discussion to *directed graphs* where the edges between nodes are referred to as *arcs*.

Using a more formal description, we say that a directed graph is a set of nodes and a set of arcs. The word *set* is defined in the mathematical sense—a collection of distinct items such that there are never duplicate items in the collection. In other words, a directed graph is a collection of distinct nodes and a collection of distinct arcs. An arc is defined as an ordered pair of nodes that establishes a one-way connection from the first node to the second node. If, for example, a graph contains an arc from node A to node B, we write the arc as (A, B). This denotes a connection from A to B but does not imply a connection from B to A.

We say that graph G = (V, E) means that graph G is composed of a set of nodes, V, and a set of arcs, E. As an example, consider the following formal definitions of an example graph.

V = {A,B,C,D,E}

E = {(A,E), (A,B), (B,A), (B,D), (C,E), (D,C), (E,B), (E,C), (E,D)}

G = ({A,B,C,D,E}, {(A,E), (A,B), (B,A), (B,D), (C,E), (D,C), (E,B), (E,C), (E,D)})

In this example, the set of vertices V is given as {A,B,C,D,E} and the set of arcs E is given by {(A,E), (A,B), (B,A), (B,D), (C,E), (D,C), (E,B), (E,C), (E,D)}. Graphs are commonly drawn as a diagram where the nodes are represented as discs, and the arcs are represented as arrows that emerge from one disc and point to another. Figure 7.10 illustrates how graph G might be drawn in this manner. Each of the five discs is labeled with one of the node names, and each arc is shown as an arrow that emerges from the first node in the arc and points to the second node in the arc.

Graphs are extremely useful models for reasoning about many different types of real-world problems. Figure 7.11 shows how the graph of Figure 7.10 can be understood as a model of airline routing. If the nodes of a graph are understood to represent airports and the arcs to represent

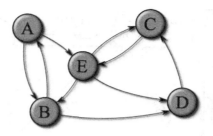

FIGURE 7.10   Graph G is drawn as a diagram. G is composed of five nodes and nine arcs.

flights from one airport to another we can represent airline connections. For the graph of Figure 7.11, we might associate node A with Seattle, B with Los Angeles, E with Denver, C with Chicago, and D with Miami. The arcs of the graph then indicate that there are direct flights from Seattle to Los Angeles and Denver but no direct flight from Denver to Seattle. Also, the graph tells us that if we want to fly round-trip from Denver to Miami we have to take a return flight through Chicago.

Road maps can be modeled as a graph if we associate each node with an intersection and each arc with one lane of a road. Figure 7.12 shows how the graph of Figure 7.10 could be interpreted as a road map. In this map, 7th Avenue is a two-way street that intersects two one-way streets: W. Oak

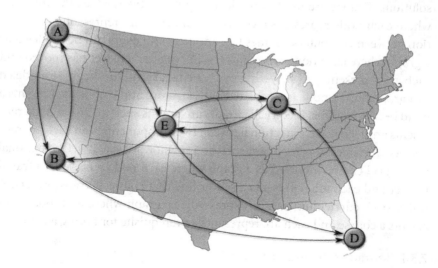

FIGURE 7.11   A graph that models airline connections between five cities in the continental United States.

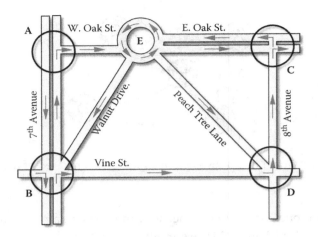

FIGURE 7.12 A graph used to model a road map. Each node represents an intersection and each arc represents a road between intersections.

and Vine. Node A represents the intersection of W. Oak and 7th while node B represents the intersection of 7th and Walnut Drive in addition to the intersection of 7th and Vine. Node E is a roundabout that gives access to W. Oak, E. Oak, Walnut Drive, and Peach Tree Lane.

Graphs are truly ubiquitous in computational thought because they are able to capture the essence of a wide variety of real-world problems and their solutions. Graphs are used to model computer networks (i.e., the Internet) where each node represents a computer and each arc represents a connection between computers. Games like checkers and chess can be modeled as a graph where each node represents the board after a player has moved and each arc represents one player's move. Chemical structures can be modeled as a graph if each node is associated with an atom and each arc represents a bond between atoms. Electrical circuits are graphs such that each node represents an electrical connection between two components, and each arc represents an electrical component such as a resistor or capacitor. The national power grid can be modeled as a graph where each node represents a transformer and each arc represents a power line that connects transformers. A university's curriculum can be expressed as a graph where each node represents a course and each arc represents a prerequisite for taking a course.

7.3.1 Terminology and Properties

Graphs are a mathematical model used to describe the central essence of a particular real-world concept or object. Graphs are often characterized

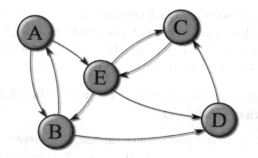

FIGURE 7.13   Graph G.

by a set of features that describe the overall structure of the graph itself. For our text, the features of most concern are listed next and illustrated by reference to graph G of Figure 7.13.

- *Adjacency*—Assume that U and V are vertices in some graph. Vertex U is adjacent to vertex V if there is an arc (U, V) in the graph. For example, vertex D is adjacent to vertex C in G since the arc (D, C) appears in G. Vertex C is not adjacent to D since the arc (C, D) is not in G.

- *Loop*—A loop is any arc such that the first and second nodes of the arc are the same. Graph G does not have a loop.

- *In-degree*—The in-degree of a vertex V is the number of arcs in the graph having V as the second vertex. The in-degree of vertex E is 2 since G has arcs (A, E) and (C, E) as the only arcs where vertex E is the second vertex.

- *Out-degree*—The out-degree of a vertex V is the number of arcs in the graph having V as the first vertex. The out-degree of vertex E is 3 since G has arcs (E, B), (E, C), and (E, D) as the only arcs where vertex E is the first vertex.

- *Order*—The order of a graph is the number of vertices. The order of G is 5 since there are 5 vertices.

- *Size*—The size of a graph is the number of arcs. The size of G is 9 since there are 9 arcs.

- *Path*—A path is a sequence of vertices such that for every pair of adjacent vertices in the sequence there is a corresponding arc in the graph. Also, a sequence containing a single vertex is a path. For example, the sequence [A, E, C] is a path since (A, E) and (E, C) are arcs in the graph. The sequence [A, B, E] is not a sequence since (B, E) is not an arc in the graph.

- *Path length*—The length of a path is the number of arcs in the path. The length of [A, E, C] is 2.

- *Cycle*—A cycle is a path where the length is greater than zero, and the first and last vertices are the same. A graph without any cycles is known as an acyclic graph. For example, [A, B, A] is a cycle and hence graph G is not acyclic.

7.3.2 Storage

Although graphs can be stored in memory using array-like techniques, we will restrict our discussion to how a graph can be represented using links. A list can be thought of as a graph where each element but the first has an in-degree of one, and each element but the last has an out-degree of one. The central difference between graphs and lists is that the vertices of a graph may have a larger out-degree.

When storing lists using a linked representation, we assumed that the out-degree of each element was one. In other words, we assumed that each element in the list was immediately followed in memory by a single link; the memory address of the single element that followed. We cannot make this assumption with graphs since a vertex may be adjacent to as many other vertices as exist in the graph. Since we cannot assume that a vertex has out-degree 1, we must store the out-degree of each vertex along with the vertex value.

We can store a graph by first storing the value of the vertex (the name) and then storing the out-degree of the vertex in the very next memory location. After this, we store N links where N is the out-degree of the vertex. As an example, consider Figure 7.14 that shows one way to store the graph of Figure 7.13 in memory. In this figure, we arbitrarily select vertex A as the anchor node and note that the anchor of the graph is then the memory location of A. In other words, the graph is anchored at 10.

Consider vertex A of Figure 7.14. Since the number 2 is stored in the memory address immediately following node A, we now know that the out-degree of A is 2. We then understand that the next two values in memory are links to the two vertices that are adjacent to A. In this example, we note that vertices B (stored at location 1) and E (stored at location 14) are adjacent to A. Note that vertex E has an out-degree of 3 and is adjacent to

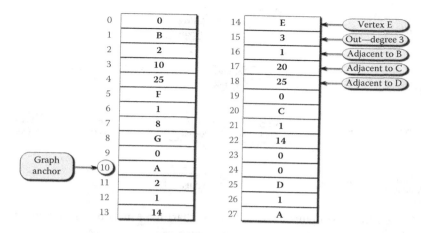

FIGURE 7.14   Graph G is stored in linear memory.

nodes B, C, and D. Note that this representation can only be used if all vertices in the graph can be reached by starting at the anchor and following some chain of arcs.

7.4 HIERARCHIES

A hierarchy is an arrangement of elements such that the elements are arranged in levels. Each element in the hierarchy may have many elements that are directly below but only one element that is directly above. Many real-world concepts form hierarchies of elements. We will briefly describe several real-world examples of data that is hierarchical in nature.

7.4.1 Organizational Chart

An *organizational chart* (often known as an *org chart*) is a diagram showing the authority structure of an organization and the relationships and reporting lines that exist between the people that are part of the organization. An organizational chart is a hierarchy of elements where the levels in the hierarchy correspond to various managerial levels, and each element corresponds to an individual within the organization.

Figure 7.15, for example, is a simplified organizational chart for the Internet sales company Amazon. In this chart, there is one president, Jeffrey Bezos, who sits atop the tree as the "root." Four individuals hold the role of vice president and report directly to Jeffrey Bezos. There are four directors and one senior manager. Since Thomas Szkutak is above Tim Stone in this chart, we understand that Tim Stone reports directly to

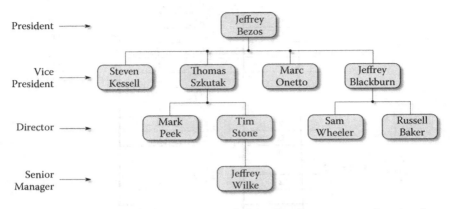

FIGURE 7.15 A simplified organizational chart for Amazon.com showing four levels of authority within the corporate structure.

Thomas Szkutak and that Thomas Szkutak has supervisory authority over Tim Stone. In addition, since Jeffrey Bezos is above Thomas Szkutak, we understand that Thomas Szkutak reports directly to Jeffrey Bezos and that Jeffrey Bezos has supervisory authority over Thomas Szkutak.

7.4.2 Family Tree

A *family tree* shows the genealogical relationship between ancestors and their descendants. The lower levels of a family tree contain family members from recent generations, and the upper levels of the hierarchy contain ancestors from many generations past. Family trees are not purely hierarchical if they include information about both the maternal and fraternal ancestry since a person will have two individuals, both a father and a mother, directly above them. This feature violates the constraint that each element in the hierarchy must have only one element that is directly above. If, however, we restrict a family tree to showing only the fraternal or maternal ancestry of a person, the result is a proper hierarchy of family members.

J.R.R. Tolkien, author of the popular *Lord of the Rings* trilogy [3], described the family ancestry of the Bagginses covering a span of about five generations. The Baggins family lived in the Shire, near the town of Hobbiton where Bilbo Baggins lived with his adopted relative Frodo Baggins. The Baggins clan is said to descend from one Balbo Baggins. Bilbo is a great-grandson of Balbo, as was Frodo's father, Drogo. Figure 7.16 shows the fraternal side of the tree for the famous Baggins family as Tolkien narrated.

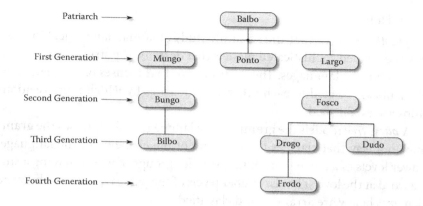

FIGURE 7.16 A simplified family tree showing the ancestry of Bilbo and Frodo Baggins.

7.4.3 Biology

Biology is the study of life and living organisms. To better understand the various forms of life, biologists group and classify living creatures according to their similarity. This biological classification is a method of scientific *taxonomy* used to group and categorize organisms into groups such as kingdom, phylum, genus, or species. Each group occurs at some level in relation to both higher and lower groupings.

Figure 7.17 shows a very small example of the taxonomy for four types of birds: bald eagles, sparrowhawks, red-tailed hawks, and buzzards. This taxonomy shows that the elements at the lowest level are species. The sparrowhawk is of the *Accipiter* genus, whereas both the red-tailed hawk and buzzard of the *Buteo* genus. The bald eagle, as shown in this taxonomy, does not possess a genus but is of the subfamily known as Haliaetus. The *Accipiter* and *Buteo* are members of the Accipitrinae subfamily while every one of these species of the same family, Accipitridae.

FIGURE 7.17 Taxonomy for a group of four bird species.

7.4.4 Linguistics

Linguistics is the formal and scientific study of human language. Linguists have developed sophisticated models that describe the structure, or grammar, of human languages. The study of grammar focuses on how language elements are related to each other and the rules by which these relationships are established.

A *parse tree* models the grammar of a language and expresses the grammar in a way that clarifies the meaning of the elements in the language. Lower levels of a parse tree contain smaller groups of words, having a single word at the lowest levels. Higher levels of the parse tree express how the elements below are arranged and classified.

An example of a parse tree is shown in Figure 7.18 where the words in the sentence "The educated student has learned computational thinking" have been grouped together into meaningful phrases. For example, the word "computational" is classified as an adjective and the word "thinking" is classified as a noun. The two-word phrase "computational thinking" is then classified as a noun phrase (as any adjective noun sequence can be classified), which is part of the larger verb phrase "learned computational thinking," which itself is part of an even larger verb phrase. At the highest level of the parse tree we find that the sequence of words "The educated student has learned computational thinking" forms a sentence.

Parse trees play a vital role not only in the linguistics of human languages, but also in the area of programming languages. When a computer

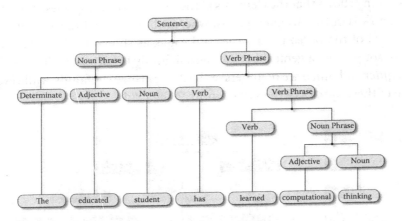

FIGURE 7.18 The structure of an English sentence is shown in this parse tree. The sentence is composed of one noun phrase and one verb phrase. Each of those phrases is composed of other grammatical elements.

programmer writes a program, the program is expressed in a computer programming language where the properties and meaning of the program are understood by creating a parse tree of the program.

As an interesting side note of historical importance, a number of important advances in the field of computer programming languages were made by a linguist named Noam Chomsky. Chomsky began his academic career by studying philosophy and later earned a Ph.D. in linguistics, an area of study that might initially seem far removed from computation. Nonetheless, Chomsky is highly regarded among computer scientists for his discoveries in the area of computer programming languages. Chomsky is credited with the development of the Chomsky hierarchy, a rigorous mathematical model of grammar, and also of the universal grammar theory. Both of these theories are vital to the field of computer science and greatly advanced the field of programming language theory.

7.4.5 Trees

A *tree* is a type of graph that is intended to model hierarchal data. More specifically, a tree is a graph that has the following characteristics:

- Exactly one vertex with in-degree zero*; this vertex is known as the root.

- Every vertex other than the root has an in-degree of one.

- There is a path from the root to every other vertex.

Figure 7.19 gives an example of a tree. The tree is a graph composed of eight vertices and seven arcs. Vertex R is the only vertex that has in-degree zero and all other vertices have in-degree one. In addition, there is a path from R to every other vertex in the graph.

While there is only one *root* node in a tree, there may be many leaf nodes. Any node that has an out-degree of zero is known as a *leaf*. There are four leaf nodes in the tree of Figure 7.19: T, E, C, and A. You might be surprised to see that the root of the tree is drawn at the top while other nodes are drawn below the root. And oddly enough, the leaves are drawn at the very bottom of the tree. Trees are conventionally drawn upside down since the ordering of the nodes implies a ranking such that the root is the node of highest importance while the leaves are nodes of least importance. Figure 7.19 illustrates this drawing convention.

---

* A tree may have an order of zero (no vertices) in which case the tree is said to be empty. In this case, there is no root vertex.

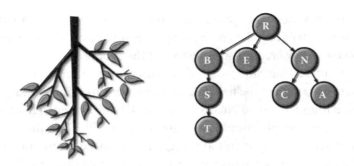

FIGURE 7.19 Trees are conventionally drawn upside down with the root at the top and the leaves at the bottom.

REFERENCES

1. Geisel, Theodore (Dr. Seuss), "Too Many Daves" from *The Sneetches and Other Stories* (New York: Random House, 1961). Dr. Seuss Enterprises, L.P.
2. Peers, Alexandra (January 2012). "Qatar Purchases Cézanne's *The Card Players* for More Than $250 Million, Highest Price Ever for a Work of Art." *Vanity Fair*. Retrieved December 18, 2013.
3. Tolkien, J. R. R. *Lord of the Rings*. (London: Allen & Unwin, 1954)

TERMINOLOGY

adjacency	indexing
anchor	leaf
arc	linked list
array	linguistics
base address	list
cycle	location
directed graph	loop
disc	memory
edge	node
family tree	order
geocaching	organizational chart
in-degree	out-degree

parse tree	taxonomy
path	track
path length	tree
root	URL
sector	word
set	zero indexing
size	

EXERCISES

1. Consider the list of numbers 22, 31, 28, 25, and 22. Show how this list could be stored in the following chunk of memory. Note that the following figure shows two columns due to space and layout considerations but you should interpret the memory as purely linear. The anchor of the list must be 26.

Address	Value	Address	Value
20		36	
21		37	
22		38	
23		39	
24		40	
25		41	
26		42	
27		43	
28		44	
29		45	
30		46	
31		47	
32		48	
33		49	
34		50	
35			

   a. The list is stored as an array using the technique described in this chapter.

   b. The list is stored using the linked technique described in this chapter.

2. Show how the graph given in the following diagram could be stored in the chunk of memory shown in Exercise 1. Ensure that the anchor of the graph is at 20. Note that there is not a unique solution to this problem.

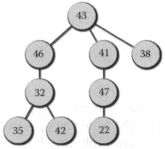

3. Draw the graph given as G = (V, E), where V = {A, B, C, D, E} and E = {(A, D), (B, E), (C, E), (D, A), (A, B), (E, A)}.

4. Show how a graph is able to model the data of a social networking site like Facebook where each user is connected to a number of other users. In this example, we will model the one-way connection that most social networking sites describe as *liking* or *subscribing*. This relationship is one-way. If, for example, user Janet decides to *like* the user Fred, this would not imply that Fred therefore also *likes* Janet.

   a. Draw a sample graph of the liking connections in a fictional social networking site. Your graph must have at least five users and at least nine *likes* connections.

   b. Does your graph have a cycle? If so, write down the cycle.

   c. What is the order and size of your graph?

5. Consider a range of memory cells containing the data shown in the following table. Draw the tree stored in memory as shown below where the anchor of the tree is 28 and the tree is organized in memory following the technique described in this chapter. Note that deciphering the structure of the tree is made somewhat difficult since the values of each node are integer numbers.

Address	Value
20	34
21	3
22	36
23	43
24	47
25	25
26	1
27	32
28	38
29	2
30	25
31	20
32	37
33	0
34	30
35	0

Address	Value
36	46
37	2
38	40
39	34
40	39
41	0
42	32
43	33
44	0
45	45
46	41
47	42
48	0
49	39
50	3

6. Consider the number 1/3. The decimal representation of this value contains an infinite number of digits since it is written 0.333333..., where the digit 3 endlessly repeats. Any decimal number that contains an endless repetition of digits is known as a repeating decimal number. Examples include 9/11 (.818181...) and 22/7 (3.142857142857...).

   a. Describe how a linked list can store a repeating decimal even though there are an infinite number of digits in the representation.

   b. Show how the number 9/11 can be stored in the memory chunk of Exercise 1 using the technique you described in part (a).

7. Both linguists and computer scientists study languages. A linguist studies human language and a computer scientist studies computer programming languages. Both disciplines make use of precise rules that attempt to express the grammatical structure of a language. Listed next are the grammatical rules of the fictional Glewphite language. Draw a tree, a syntax tree, that describes the grammatical structure of the Glewphite sentence: "loza glo glargle le argl."

   • The following items are Glewphite nouns.

     – argl

     – glo

     – zargl

- The following items are Glewphite adjectives.
  - le
  - gleglom
- The following items are Glewphite verbs.
  - glargle
  - loza
- A Glewphite sentence is a sequence of phrases. Each of the following is a Glewphite phrase.
  - A verb followed by an adjective followed by a noun
  - A noun followed by an adjective
  - A verb followed by a noun

8. Draw the family tree of one of your grandparents. Choose to draw either the maternal or fraternal elements of the family tree. The tree must have at least three generations and at least three individuals in each generation. If you do not know the details, include as much information as you know; and invent fictional family members to round out your family tree. When you are done, show how the family tree could be stored in the chunk of memory given in Exercise 1.

9. Give the formal definition for the road map graph described in Figure 7.12. In other words, give a set of nodes and a set of arcs that define the map.

CHAPTER 8

Algorithmic Thinking

I have no magic formula. The only way I know to win is through hard work.

—DON SHULA

OBJECTIVES

- To understand the von Neumann principle
- To understand the concept of two-dimensional data layout and tabular information retrieval
- To understand the relationship of discrete functions and tables
- To understand how formulas can be used for textual processing
- To understand how formula can be used to define and process patterns

---

From a computational viewpoint, a formula not only describes a process for computing a value but can also be understood as the value itself. This chapter first describes how formulas are used in the context of spreadsheets. Numerical operators are characterized in terms of arity, associativity, and precedence. In addition, the formulas of a spreadsheet are described as a model that specifies the relationship between the input values and the output values generated by the spreadsheet.

We later show that formulas are able to process not only numbers but also textual data. Text processing is described apart from spreadsheets using powerful formulas that create text, examine parts of text, and define textual patterns. This chapter shows how textual formula can be used to study DNA and to potentially improve Internet searches.

8.1 VON NEUMANN ARCHITECTURE

The earliest computing machines were designed with a *fixed-program architecture*. A fixed-program architecture implies that the instructions for the computer are built into the actual hardware of the machine itself, and only the data that is processed by the computer is stored in memory. These computers were not truly reprogrammable since changing the program required that the machine itself be rebuilt or rewired to follow a new sequence of instructions.

Certain modern computing systems can be understood as using a fixed-program architecture. A basic hand calculator, for example, can perform a small number of arithmetic tasks, but it cannot be programmed for text processing or instant messaging. Many elements of a graphics system also follow a fixed-program architecture since graphical processing is computationally expensive and can be accelerated by specialized hardware.

In contrast to a fixed program architecture, a *stored-program architecture* is one that stores both program instructions and program data in memory. John von Neumann, one of the greatest mathematicians in recent world history, is often given credit with the development of the stored-program architecture although his work was based on the earlier efforts of Allen Turing. The term *von Neumann architecture* has become synonymous with stored-program architecture due to von Neumann's publication of a 1945 article titled "First Draft of a Report on the EDVAC" [1]. This technical article described a computer with a processing unit, a control unit containing an instruction register and program counter, external memory, and internal memory to store both data and instructions.

The von Neumann architecture allowed computers to treat both program instructions and program data in a nearly identical manner. Since a program's data could be changed by a program, it followed that a program could change its own instructions. It also followed that if a program could generate data as output, a program could also generate a set of instructions as output. In other words, the von Neumann architecture allowed programs to reprogram themselves; a feature that is commonly known as *self-modifying code*. Over time, computer programmers discovered that the ability to treat instructions as data gave them incredible power. Many of the programs that are in use today have been produced with the aid of assemblers, compilers, and other computational programming tools.

These tools are essentially programs that write programs, programs that treat the code as data.

8.2 SPREADSHEETS

A *spreadsheet* is an interactive program designed to organize, process, and display data in a tabular form. Although spreadsheets were initially developed to support financial bookkeeping and accounting efforts, they are commonly used in any domain where tabular data is stored and processed. We are motivated to describe spreadsheets as an example of the duality that exists between code and data. We specifically describe how formulas, the code of a spreadsheet, produce data and how a spreadsheet often treats the code as the data.

8.2.1 Spreadsheet Structure

A single *worksheet* is a tabular arrangement of cells such that a single cell is located at the junction of a row and column of the sheet. A *cell* is a repository, or a box, that contains data. Each cell has a unique name and roughly corresponds to the concept of a variable in a programming language since data can be placed into the cell and the data can be later accessed through the name of the containing cell.

The rows of a spreadsheet are typically numbered starting from 1 and the columns are labeled alphabetically starting from the letter A. A cell is named with respect to its location in the spreadsheet. The name of a cell is formed by joining the column and the row of the cell. A cell that is located in column C and row 4, for example, is named C4. Figure 8.1 shows this structure.

Each cell of the spreadsheet can contain a number, text, or formula. Formulas are used to derive new data from data that already exists, perhaps as the result of some other formulaic output, elsewhere in the spreadsheet. If a change is made to any individual cell, all other cells in the spreadsheet

	A	B	C	D	E
1					
2					
3					
4					
5					

FIGURE 8.1   Spreadsheet structure. The highlighted cell occurs in column C and row 4. The name of the cell is C4.

will be updated to ensure that all cells in the spreadsheet are internally consistent. In other words, any portion of the spreadsheet that could be affected by the change to a cell will be automatically updated to reflect that change. This automatic updating feature makes a spreadsheet useful for what-if analysis, since users can change input values in the spreadsheet and quickly visualize how these changes affect the output values.

Any cell that contains a formula will display the data computed by the formula and not the formula itself. Although spreadsheet users must be aware of formulas when they create a spreadsheet, the spreadsheet user will normally be more concerned with the values produced by the formulas and hence the spreadsheet displayed only what the formulas produce.

8.2.2 Formulas/Expressions

Formulas are the primary computational element of a spreadsheet. A *formula* defines the computation required to take input data and produce an output datum. Computer programmers also refer to formulas as expressions. In this section we will use the term *formula*. The ability to write correct and meaningful formulas requires a solid understanding of every element that might occur in a formula. In this section we describe these elements and carefully define their computational meaning.

Most spreadsheet applications require that an equal symbol (=) be typed prior to the formula. The equal symbol is merely a way of notifying the spreadsheet that the cell contains a formula.

All formulas are required to follow a rigid internal structure or syntax. This structure enables the computer to precisely determine the computational sequence of steps that will generate an output value from some set of inputs. In particular, formulas may consist only of the following elements:

- *Numbers*—3, −18, and 5.3 are examples.

- *Operators*—+, −, *, and / are examples.

- *Cell references*—A1 and C4 are examples.

- *Functions*—MAX() and MIN() are examples.

8.2.2.1 Numbers

Numbers are the simplest elements of a formula. Other formula elements may process numbers and produce numbers as output. These other formula elements are described in the following subsections.

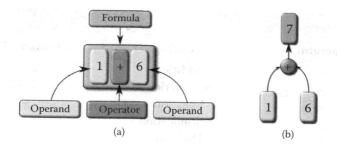

FIGURE 8.2 (a) The structure of a formula involving the addition operator. (b) The operator takes the numbers 6 and 1 as input and produces 7 as the output.

8.2.2.2 Operators

In computer programming, an *operator* is a symbol that represents a computational action. Examples include the plus symbol (+) that represents mathematical addition, and the asterisk symbol (*) that represents multiplication. Every operator accepts one or more inputs, also known as *operands*, and generates exactly one output value. The operands are the inputs to the operator in order to generate an output value. Figure 8.2 gives an example of a formula that involves the use of the addition operator. There are two operands to the addition operator, the number 1 and the number 6. The addition operator is denoted by the plus symbol (+) while the entire phrase is properly understood as the formula. The output of the operator is the number 7 and, since there is only one operator in this formula, the output of the entire formula is also 7.

Arithmetic operators perform basic numeric operations such as addition and subtraction, whereas comparison operators compare two values. A brief summary of common operators used in a spreadsheet is given in Figure 8.3.

Formulas will often contain multiple operators such that the output of one operator becomes the input, or operand, to another. Consider, for example, the formula of Figure 8.4 that involves both an addition and multiplication operator. In this example, the output of the multiplication operator becomes the input to the addition operator.

Formulas calculate values in a specific order and the result of a formula with more than one operator is dependent upon the order in which the operands are applied. If, for example, we perform the addition of Figure 8.4 before multiplication, the resulting value is 9. On the other hand, if we perform multiplication before addition, we obtain the value 7.

Common Arithmetic Operators		
Operator	Meaning	Example of Use
+	Addition	2 + 3
−	Subtraction	2 − 3
−	Negation	−3
*	Multiplication	2 * 3
/	Division	2 / 3
^	Exponentiation	2 ^ 3

Common Comparison Operators		
Operator	Meaning	Example of Use
<	Less than	2 < 3
<=	Less than or equal to	2 <= 3
=	Equal to	2 = 3
<>	Not equal to	2 <> 3
>	Greater than	2 > 3
>=	Greater than or equal to	2 >= 3

FIGURE 8.3 Common operators that occur in spreadsheet formulas.

The order in which the operators are calculated can be explained by visualizing how the various operators are interrelated. Consider, for example, the number 2 in Figure 8.4. This number might be understood as the second operand of the addition operator or it might instead be understood as the first operand of the multiplication operator. It cannot be both since this would essentially break the formula into two independent pieces: 1+2 and 2*3.

In a logical sense, therefore, there are only two possible interpretations, or results, for this formula. The first interpretation is to take the values 1 and 2 as the operands of the addition operator and take the output of the

$$1 + 2 * 3$$

FIGURE 8.4 A formula involving two operators. The output of one operator is the input to another.

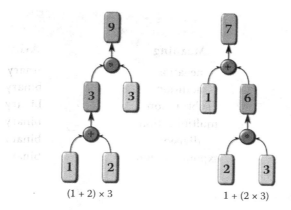

FIGURE 8.5   Two distinct interpretations of the formula $1 + 2 \times 3$.

addition operator as the first operand of the multiplication. This interpretation applies the addition operator before the multiplication operator, as is shown on the left of Figure 8.5. The second interpretation is to take the values 2 and 3 as operands of the multiplication operator while the value 1 and the multiplicative output serve as the operands of the addition operator. This second interpretation applies the multiplication operator before the addition operator, as is shown on the right of Figure 8.5.

Since a formula cannot simultaneously generate two different results, the computer must be able to determine the order in which the operators of a formula will be applied. The correct ordering of operator application can be determined by understanding three distinct properties for each of the operators. These three properties are the *arity, precedence,* and *associativity.* Understanding these three properties allows the computer to determine how to correctly interpret any formula that involves more than one operator.

8.2.2.2.1 Arity

The *arity* of an operator is the number of inputs required by the operator. A unary operator is an operator that is of arity 1, an operator requiring one input. A binary operator is an operator that is of arity 2, an operator requiring two inputs. A ternary operator requires three inputs. Addition, subtraction, multiplication and division are examples of binary operators, and arithmetic negation is an example of a unary operator. Arithmetic negation is defined as subtraction except that the first operand is defined to be 0. This implies that when we write a phrase such as "−5", the negation operator understands this to mean "0−5". Most spreadsheet applications

Arity of Operators		
Operator	**Meaning**	**Arity**
−	negation	unary
+	addition	binary
−	subtraction	binary
*	multiplication	binary
/	division	binary
^	exponentiation	binary

FIGURE 8.6   Arity of operators.

and many programming languages do not include ternary operators. Figure 8.6 lists the arity of each arithmetic operator.

8.2.2.2.2 Precedence

Precedence defines the relative order in which operators are applied. In any formula, the next evaluated operator is always the highest precedence operator that has not yet been evaluated (unless parentheses are used to group elements together). All higher precedence operators are evaluated before any lower precedence operators. The precedence of common spreadsheet operators is given in Figure 8.7. This table indicates that negation is performed before any other operator, and the comparison operators are performed last in any formula. If a formula contains two different operators having the same precedence, we need a tie-breaking rule to determine the order of application. We break ties by saying that if a formula contains two different operators of identical precedence, then the operators are evaluated in left-to-right order.

Operator Precedence	
Operator	**Meaning**
−	Negation
^	Exponentiation
*, /	Multiplication and Division
+, −	Addition and Subtraction
<, <=, =, <>, >=, >	Less than, less than or equal to, equal to, not equal to, greater than or equal to, greater than

FIGURE 8.7   Operator precedence table.

$$1 - 2 - 3$$

FIGURE 8.8 Associativity determines the ordering of multiple identical operators in a formula.

8.2.2.2.3 Associativity

Associativity determines the order in which operators are applied if all the operators in a formula are identical. Consider, for example, Figure 8.8 where there are two subtraction operators.

Left-associative operators are ordered from left to right, and *right-associative* operators are ordered from right to left. In most spreadsheet applications, the operators are all left associative. Other programming languages, however, recognize that there are operators that should be understood as intrinsically right associative. Exponentiation, for example, is a right-associative operator in many languages.

Knowing the associativity of an operator is necessary to correctly understand any formula that contains that operator. Consider the formula of Figure 8.8. If we believe that subtraction is right associative, we believe that the first computational action performed by the computer is to subtract 3 from 2 and produce the value –1. The last computational action is to subtract –1 from 1 and produce the value 2 as a result! If, however, we correctly understand that subtraction is left associative, we know that the first computational action is to subtract 2 from 1 and produce the value –1. The last computational action is to subtract 3 from –1 to produce the value –4. These two interpretations are illustrated in Figure 8.9.

8.2.2.2.4 Evaluation Example

Figure 8.10 shows a step-by-step example of how the operators in a complex formula are evaluated. In this example, there are five operators: addition, exponentiation, multiplication, subtraction, and division. As a result, there are five distinct steps required to compute the result of the

$$1 - (2 - 3) \qquad\qquad (1 - 2) - 3$$

right-associative interpretation   left-associative interpretation

FIGURE 8.9 The associativity of an operator determines the meaning of the operator.

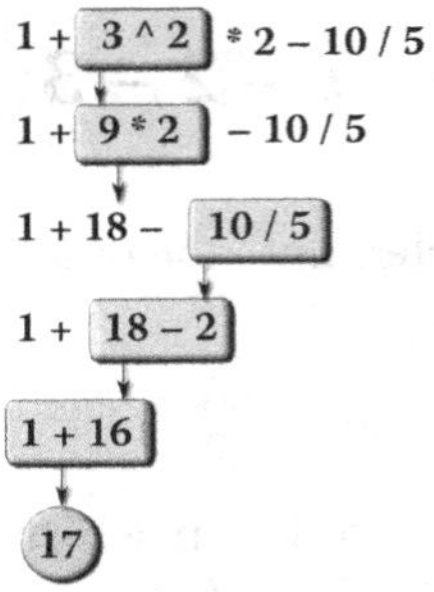

$$1 + \boxed{3 \wedge 2} * 2 - 10 \, / \, 5$$
$$1 + \boxed{9 * 2} - 10 \, / \, 5$$
$$1 + 18 - \boxed{10 \, / \, 5}$$
$$1 + \boxed{18 - 2}$$
$$\boxed{1 + 16}$$
$$17$$

FIGURE 8.10 Evaluating a formula that involves more than one operator. The order of the operators is determined by the precedence, associativity, and arity of the unevaluated operators in a formula.

formula. Each action in this figure is denoted by an arrow connecting one line to the next in the chain of evaluations. The specific formula elements involved with each action are highlighted at each stage of computation.

Exponentiation is the first operator that the computer evaluates since it is the highest precedence operator. This operator is applied to the operands 3 and 2 and calculates 9 as the output produced by the subformula 3^2. Since code and data are effectively interchangeable, we can replace the code 3^2 with the output data 9.

Multiplication is the next operator that the computer evaluates. Although multiplication and division have the same precedence, we choose multiplication over division because our tie-breaking rule says that when two different operators have the same highest precedence, we evaluate the operators in left-to-right order. Since multiplication is to the left of division, we choose to evaluate multiplication before division. Since the produce of 9 and 2 is 18, we can replace the formula 9 * 2 with the output data 18.

The third operator is division since it has the highest precedence of any of the remaining operators. We divide 10 by 5 to obtain 2. The value 2 is then substituted for the corresponding formula. After division, we perform subtraction followed by the final operator: addition. Again, although addition and subtraction have the same precedence level, we move through these operations in left-to-right order. The output of the addition operator is also the output of the formula since it is the last operator to be evaluated. In the example of Figure 8.10, the formula "1 + 3^2 * 2 – 10/5" expresses a process for generating the number 17.

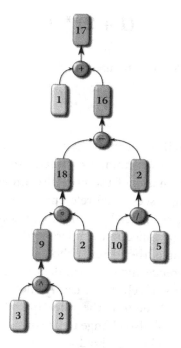

FIGURE 8.11 A formula structure.

The process of evaluating a formula is often expressed as a tree structure as shown in Figure 8.11. The tree structure is useful because it explicitly indicates how the operators are related to each other and how the operands are fed into each operator. The ordering of the operators is also explicitly defined; the ordering proceeds from bottom to top and left to right. Each operator is drawn as a blue circle where the inputs are denoted using arrows that flow into the circle. The operands that occur in the formula are colored yellow and the values that are produced by operators are colored green.

8.2.2.2.5 Parentheses

Parentheses can be used to change the order in which operators are evaluated. Specifically, any part of a formula that is enclosed in parenthesis can be thought of as an independent part of the whole that has precedence above all others. For example, the formula of Figure 8.12 produces 9 rather than 7 because the computer first evaluates the subformula 1 + 3 since the parentheses isolate that group from the whole. The computer finally multiplies 3 by 3 to obtain 9.

$$(1 + 2) * 3$$

FIGURE 8.12 Parentheses can be used to change the order of evaluation by grouping subexpressions.

8.2.2.3 Cell References

A spreadsheet cell can be understood as a box that contains valuable data. Since each of these boxes is uniquely named, we can access the data in the box by giving the name of the box, also known as a *cell reference*. Whenever the computer sees a cell reference in a formula, the computer gets the data from the referenced cell. In other words, the value of the cell reference is the data contained in the referenced cell.

Consider, for example, the formula shown in Figure 8.13. In this formula, A2 is a cell reference and the formula produces a value by multiplying 3 with the contents of cell A2. If cell A2 holds the value 2, then this formula produces the value 6. If the value in cell A2 later changes, the value of this formula will also change to be consistent with the new value held in A2. Cell A2 might even hold a formula. If so, the result of that formula is then multiplied by 3 to determine the output of the formula.

Figure 8.14 illustrates how cell references can be used to bring data into a formula. In this spreadsheet, Amy, Sara, and Jim are classmates in a local high school and they all use the same social networking website to connect with their friends. Amy has 23 friends in this network and Sara has 10 friends and Jim has 15 friends. These three numbers are the input to the spreadsheet and they are placed into cells A2, B2, and C2.

These three classmates want to compute the total number of friends that they, as a group, are connected to. They therefore enter "=A2+B2+C2" into cell D2 of the spreadsheet. They understand that cell D2 computes one value by summing the contents of cells A2, B2, and C2. They also want to know the average number of connected friends and so they divide their total number friends by three. They enter "= D2/3" into cell E2. This formula divides the total number of friends by 3 to obtain the average.

$$3 * A2$$

FIGURE 8.13 A formula containing a cell reference.

	A	B	C	D	E
1	**Amy**	**Sara**	**Jim**	**Total**	**Avg**
2	23	10	15	=A2+B2+C2	=D2/3

FIGURE 8.14 A spreadsheet to determine the total and average number of friends that Amy, Sara, and Jim have on a social networking site.

The formula in cell D2 of Figure 8.14 tells the computer to take the numbers contained in cells A2, B2, and C2, then add them together. In this example, the formula is then reduced to = 23+10+15 since these are the values contained in the referenced cells. Figure 8.15 shows how the data flows into the formula of cell D2 and that the value produced by the formula is 48. The result of formula D2 then flows into the formula of E2. Also note that while cell D2 contains a formula, the spreadsheet does not display the formula, but rather displays the number 48, the output data.

The formula in cell E2 of Figure 8.15 tells the computer to take the data contained in cell D2, then divide it by 3. Although we might initially believe that this cannot be done since cell D2 contains a formula and not a number, the computer recognizes that the formula produces a number; the code is data. Hence, the number produced by the formula of D2 is divided by 3 to obtain 16 as the average number of friends the three classmates connect with.

Most formulas depend upon other cells of the spreadsheet and hence a spreadsheet must determine the order in which formulas should be evaluated. The formula of E2, for example, cannot be computed until the formula for D2 has produced a value. Spreadsheets determine the order in which formula must be evaluated by analysis of the *data dependencies* of the spreadsheet. A formula is said to depend upon another cell if the formula contains a reference to the cell. Figure 8.14 shows that cells A2, B2,

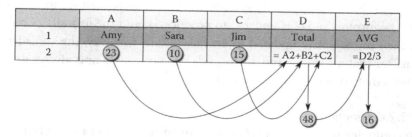

FIGURE 8.15 Illustration of the flow of data through the spreadsheet.

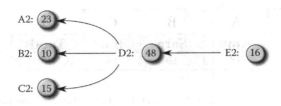

FIGURE 8.16 The order in which formulas are evaluated is determined by the data dependencies.

and C2 are not dependent upon any other cell since these cells contain no formula. Cell D2, however, is dependent on A2, B2, and C2 since these cells are referenced by the formula of D2. Using similar reasoning we see that E2 depends on cell D2 since D2 occurs in the formula. Also note that since E2 depends on D2 and since D2 depends on A2, B2, and C2, we can conclude that E2 depends on each of these other cells.

If a spreadsheet contains many formulas, the spreadsheet must first analyze the dependencies of every formula to determine which formula it should evaluate first. The spreadsheet will perform this analysis by forming a data dependency graph. A data dependency graph is a directed graph showing the dependencies that exist between cells. In a data dependency graph, each dependency is shown as an arrow that points from a cell to a cell on which it depends. Figure 8.16 shows a data dependency graph and indicates how the spreadsheet program might analyze such a graph to determine the order in which formula are evaluated. In this example, cells A2, B2, and C2 hold values and hence have no dependencies. This is reflected in the dependency graph since there are no outgoing arrows. The formula of D2 can be evaluated first since D2 depends only on cells that already hold values. Once D2 has produced a value, that value can be fed into E2, and since E2 has no other dependencies it can then be immediately evaluated.

Formulas of a spreadsheet are reevaluated whenever a dependent value changes. If, for example, the spreadsheet user changes cell A2 to the value 26, the formula of D2 is updated to produce 51, and the formula of cell E2 is then immediately updated to produce 17. In this way, the spreadsheet ensures that all values and all formula are up to date and always display values that are internally consistent.

8.2.2.4 Functions
In computer science, a *function* is a named sequence of instructions that will produce an output value when it is provided with input. Although

FIGURE 8.17   How to use a function. Type the function name and then, in parenthesis, type a list of input values where the values are separated by commas.

some functions may produce values even when no inputs are provided,
functions will almost always be given input and this input will be used
to compute the correct output value. Functions are used, or called, by
typing in the name of the function followed by a comma-separated list
of input values. These values are enclosed in parenthesis to indicate that
they are grouped together as inputs to the function. When the spreadsheet
evaluates the function, the function produces an output value. Figure 8.17
shows the syntax for correctly making use of a function.

8.2.2.4.1 Ranges

Common spreadsheet functions include *max, min, average,* and *sum*. Each
of these functions takes a range of cells as input and produces an output
that depends on the data contained in the range. In spreadsheets, a *range*
of cells is a rectangular subregion of a spreadsheet that may span many
rows and many columns. A range is denoted by first typing the upper-
left cell reference of the range, then typing a colon (:), and then typing in
the lower-right cell reference. The range denoted as A2:C3, for example,
includes the cells A2, B2, C2, A3, B3, and C3 as shown by the highlighted
cells of Figure 8.18.

Each of the max, min, average, and sum functions accepts a range and
computes an output that depends on the range. Each of these functions is
illustrated in Figure 8.19. Each function is applied to the range A2:C3 of
Figure 8.18 to produce the value in the second column. A brief description
of each function is then given in the right-most column.

	A	B	C	D
1	3	1	0	2
2	5	2	0	4
3	7	3	1	2
4	3	4	4	1

FIGURE 8.18   The range A2:C3 includes the highlighted cells.

Function Call	Value	Description of function
Max(A2:C3)	7	Computes the maximum value of all cells in the range
Min(A2:C3)	0	Computes the minimum value of all cells in the range
Sum(A2:C3)	18	Computes the sum of all cells in the range
Average(A2:C3)	3	Computes the average value of all the cells in the range

FIGURE 8.19   Example of the functions max, min, sum, and average.

Recall the case study of the three friends that used a spreadsheet to analyze their social network list of friends. Once they learn about functions, the three classmates rewrite their spreadsheet to exploit the sum and average functions. They decide to replace the formula of D2 that previously read =A2+B2+C2 with the more concise =SUM(A2:C2). The value produced by this formula is the sum of the cells in the range A2:C2. They also decide to replace the formula of E2 that previously read =D2/3 with =AVERAGE(A2:C2) since the average function produces precisely the value that the classmates need for that cell. These changes are shown in Figure 8.20.

In summary, we have used spreadsheets to illustrate the power of viewing code and data as interchangeable. We have described how computers evaluate formula by analysis of the arity, precedence, and associativity of operators in addition to the use of dependency graphs. The ideas that we have introduced extend well beyond the context of a spreadsheet. Throughout the remainder of this chapter we will describe how computers are used to process not only numbers but also text and textual patterns.

	A	B	C	D	E
1	Amy	Sara	Jim	Total	Avg
2	23	10	15	=SUM(A2:C2)	=AVERAGE(A2:C2)

FIGURE 8.20   The use of functions to compute output values.

8.3 TEXT PROCESSING

Although much of the information stored on computers involves numbers, the majority of data is textual rather than numeric. For example, your name, social security number, address, and Facebook status are all textual in nature. In this chapter we will explore how computers can be used to create, process, and reason about textual information.

Computer programmers use the term *string* when referring to textual data. A string is simply a piece of text, or more formally, an ordered sequence of individual characters. You can think of a single character as the result of any one key that you press on a keyboard. A character is usually a letter of the alphabet, but a character might also be a punctuation symbol such as a comma, semicolon, or question mark. A character might even be a nonprintable character such as a tab or a linefeed. The length of a string is the number of characters contained in the string. The length of a string may be zero, if the string contains no characters, and it may be larger than zero. No string has a negative length since it is not possible to have a sequence that contains a negative number of characters.

8.3.1 String Basics

In most programming languages, string data is denoted by using double quotes to surround the text. For example, "Hello" is a string having five characters. Any sequence of characters that is enclosed by double-quotes is known as a *string literal*. The double quotes are not part of the string itself; they simply serve to notify the computer that the enclosed text is a string literal.

Since digits appear on the keyboard, strings may contain digits in addition to alphabetic characters. Consider, for example the string literal "04/13/65". This string has a length of 8 where the characters at indices 2 and 5 are both forward slashes (/), but the remaining characters are digits. When we see this string literal we understand that this is a shorthand notation for writing a date. The string literal indicates the thirteenth day of the fourth month of the year 1965. In other words, the string literal should be understood as the date of April 13, 1965. If, however, we were to remove the double quotations from the expression, a computer would interpret the expression much differently than a sequence of characters. The computer would first divide the number 4 by the number 13 and then take the result and divide it by the number 65.

0	1	2	3	4
H	e	l	l	o

FIGURE 8.21 Each character of the string "Hello" is given an index. The indices start at 0.

The characters in a string are indexed such that the first character has an index of 0, the second character has an index of 1, and so on. If we again consider the string "Hello", we note that the character at index 0 is H, the character at index 1 is e, and the character at index 4 is o. Figure 8.21 shows how the characters of a string are indexed. The top row shows the indices of each character, while the characters themselves occur in the second row.

Note that the length of the string "Hello" is 5 while the largest index is 4. This observation suggests that for any string of length N, the largest valid index is N − 1. Since the smallest possible index is always 0, we note that for any string of length N, the only valid indices are in the interval 0 to N − 1.

8.3.2 String Operations

8.3.2.1 Indexing

Strings are often analyzed by using indices to access the individual characters of the string. Given a string literal we can access the character at a particular index by using a bracket notation. This is referred to as an *indexing* operation. Figure 8.22 shows us how we can obtain, for example, the fourth letter of the string literal "popcorn". Since the fourth character has an index of 3, we write the number 3 inside of the brackets following the string literal. This expression produces a string containing one letter,

---

Example of Indexing

"popcorn"[3]

FIGURE 8.22 Indexing can be used to obtain individual characters of a string literal.

Example of Finding the Length

"popcorn".length

FIGURE 8.23   How to obtain the length of a string. The length is an integer number.

a lowercase c. Of course, if the index we write inside of the brackets is not valid, the resulting expression will be understood as an error. For example, the expression "popcorn"[15] produces an error since the only valid indices are in the interval 0 to 6.

8.3.2.2 Length

We can obtain the *length* of a string literal by typing a ".length" after the string. Figure 8.23 shows how we can obtain the length of the string literal "popcorn". The expression "popcorn".length produces the number 7 since the length of the string literal "popcorn" is 7. In other words, the code "popcorn".length is interchangeable with the value produced by the code, the number 7.

8.3.2.3 Concatenation

String *concatenation* is another common string processing operation. String concatenation is an operation that takes two strings and splices them to form a third string as output. String concatenation is usually expressed, oddly enough, as a plus symbol (+). Although we usually think of the plus symbol as referring to the mathematical addition of two numbers, the plus symbol is also employed to concatenate two strings. Figure 8.24 shows

Example of String Concatenation

"pop" + "corn"

FIGURE 8.24   Strings can be concatenated (added) to produce a new string.

how the two strings "pop" and "corn" can be concatenated to produce the string "popcorn".

8.3.2.4 Naming

Earlier in this textbook we described how variables can be used to refer to numeric data. Recall that variables are bound to data through a name-binding operation that we denote using the left-arrow symbol (←). On the left of this symbol must be a variable name and a value must occur on the right of the arrow. Figure 8.25 shows how we might use string variables to refer to string literals. In this sequence of actions we tell the computer to (1) bind the name x to the string literal "pop", (2) bind the name y to the string literal "corn" and (3) bind the name z to the string "popcorn", a string that is produced by concatenating the strings referred to by the variables x and y.

String processing can be confusing when variables are used to refer to a string because it can be difficult to distinguish between the name of a string and the string itself. Computer scientists say that names occur in a namespace, whereas values occur in a value space. The variables x, y, and z of Figure 8.25 occur in the *namespace* of the program, while the strings "pop", "corn", and "popcorn" are in the *value space* of the program. When processing with text, it is vitally important to keep in mind the distinction between the name of a string and the value of the string.

Example of String Variables
x ← "pop"
y ← "corn"
z ← x + y

FIGURE 8.25 Name binding with strings. Variables x, y and z are names while "pop", "corn" and "popcorn" are in the value space.

Example of Substring

x ← "computational thinking"
y ← x.substring(3, 6)

FIGURE 8.26   A subsequence of a string can be produced by using the substring method.

8.3.2.5 Substring

Although indexing allows us to obtain an individual character of a string, we often want to obtain a subsequence of a string. The *substring* function allows us to obtain part of a string if we know the indices of the first and last characters that we want to extract from the string. An example of the substring function is shown in Figure 8.26. The function substring is applied to the x variable, which is bound to the string "computational thinking". We are telling the computer to give us the sequence of characters starting from the character at index 3 and ending with the character at index 5. When reading this paragraph you might initially believe that there is a mistake in the text since Figure 8.26 uses the number 6 as the second input into the substring function. The number 6 denotes the index of the first character that is not included in the output; hence the last character included in the subsequence produced by the substring method is at index 5 rather than 6. When a computer executes this code, the variable y is bound to the string "put" since this is the sequence of characters spanning indices 3 to 5 of the variable x.

8.3.2.6 Searching

Sometimes it is useful to find the index of some character in a string literal. Consider, for example, the e-mail address "elvispresley@heartbreak. hotel.com". We might want to know where the at sign, or ampersat, character (@) occurs in the string so that we can split the string into two parts: the user name and the name of the e-mail service. The *indexOf* function searches a string literal for a character and returns the index of the first occurrence of the character. Figure 8.27 shows how the string literal "popcorn" is searched to find the index of the first lowercase c. In this example, the value produced by this code is the number 3 since that first lowercase c occurs at index 3 in the string literal "popcorn". In other words, the code

Searching with IndexOf

$$x \leftarrow \text{"popcorn".indexOf("c")}$$

FIGURE 8.27   Using indexOf to find the index of the first lowercase c in "popcorn".

"popcorn".indexOf("o") is interchangeable with the number 3, the data the code produces. If we had instead searched for the first lowercase z we would have obtained the number –1; an invalid index that informs us that the character z is not contained in the string literal "popcorn".

8.3.2.7 Case Study: Processing e-Mail Addresses

We have already asserted that textual data is more prevalent in a computational system than numeric data. In this case study we will consider how e-mail addresses, a very common piece of textual data, can be automatically processed and analyzed for use in a business setting.

Perhaps we are creating a company to sell T-shirts to college students. We establish a policy that requires users to register prior to browsing our catalog and ordering products. We require that each user provide an e-mail address and a password. We know that an e-mail consists of two general parts: a user name (also referred to as the local part) and a host site (also referred to as the domain part). These two parts are separated by the ampersat symbol. Figure 8.28 shows how an e-mail address is broken into two parts by the ampersat symbol.

To verify that the user is a college student, we establish a policy requiring that the provided e-mail address terminate with the characters *edu*. This part of an e-mail address is referred to as the domain extension. By convention, any e-mail address having *edu* as the domain extensions is understood to be an educational institution. In addition we would also like to separate the user name from the e-mail host site so that we can track the number of users that are associated with each educational institution.

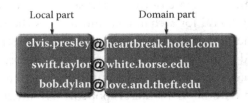

FIGURE 8.28   e-Mail addresses have a local (user name) and domain (host site) part.

e-Mail Analysis for the Beatles

address ← *readAddressFromUser()*
username ← address.substring(0, 9)
hostsite ← address.substring(10,28)
extension ← address.substring(25,28)

FIGURE 8.29 Extracting information from an e-mail address.

We realize that our web registration system must accept any e-mail address typed in by the user and then extract three vital subsequences: the user name, the host site, and the domain extension. As an example, consider the e-mail address *bob.dylan@love.and.theft.edu*. This e-mail address has a length of 28 and consists of the user name *bob.dylan*, the host site is *love.and.theft.edu*, and the final three characters of the host site are *edu*.

For this specific e-mail address, we can use the text-processing commands shown in Figure 8.29 to extract the three relevant strings. Note that the command to readAddressFromUser is simply a way of expressing that the registration web page has a text-entry field from which we obtain the text entered by the user. Since the ampersat occurs at index 9, we know that the username consists of the first 9 characters of the address and hence we use the substring statement to extract the corresponding character sequence. Also, since the ampersat is at index 9, we can extract the host site by taking the characters starting at index 10 and moving up until the end of the string. The final three characters are those characters whose indices are given as 28 − 3, 28 − 2, and 28 − 1 where 28 is the length of the address.

Although this code correctly extracts the username, host site, and extension from the e-mail address *bob.dylan@love.and.theft.edu*, it will not work for most other e-mail addresses. If, for example, we execute the code of Figure 8.29 on the e-mail address *elvis.presley@heartbreak.hotel. com* we find that the username is *elvis.pre*, the host site is *ley@heartbreak. hot*, and the extension is *hot*. Although we might believe that the phrase *hot* would make a great domain extension, there is no domain extension for *hot* websites.

The code of Figure 8.29 can be generalized to work correctly on all valid e-mail addresses. We recognize that our code makes two assumptions that are not generally true of all e-mails. Those assumptions are that (1) the ampersat occurs at index 9 and (2) the length of the e-mail address is 28.

e-Mail Analysis for Any Valid e-Mail

> address ← *readAddressFromUser()*
> ampersatIndex ← address.indexOf("@")
> length ← address.length
> username ← address.substring(0, ampersatIndex)
> hostsite ← address.substring(ampersatIndex+1, 28)
> extension ← address.substring(length-3, length)

FIGURE 8.30   Extracting information from an e-mail address.

The first assumption is encoded in the phrase "address.substring(0, 9)" since we used the number 9 as a result of assuming that the index of the ampersat is 9. The first assumption is also encoded in the phrase "address.substring(10,28)" since we understood the number 10 to be the index of the first character following the ampersat. The second assumption is also encoded in the phrase "address.substring(10,28)" since we understand the 28 to be the length of the address.

We can remove these assumptions by first finding the index of the first occurrence of the ampersat in any address and then finding the length of the address. We can then make use of the values to extract the username, host site, and extension from any e-mail address the user chooses to type. Our modified code is given in Figure 8.30.

Consider using this code to process the e-mail address *elvis.presley@ heartbreak.hotel.com.* We first use the indexOf function to determine that the ampersat occurs at index 13. We then determine that the length of the address is 34. Since the ampersat occurs at index 13, we know that the username must consist of the characters between indices 0 up through 13 (but not including the character at index 13). Again, since the ampersat occurs at index 13 and the length of the address is 34, we know that the host site must consist of the characters between indices 14 up through 34 (but not including the character at index 34). Finally, the extension must consist of the characters between indices 28-3 up through 28 (but not including the character at index 28).

8.3.2.8 Case Study: Processing Dates

The European date format is essentially the reverse of the American date format. Most Americans write a date by putting the month before the day while Europeans put the day ahead of the month. The date April 13, 1965,

e-Mail Analysis for the Beatles

edate ← *readDateFromUser()*
day ← edate.substring(0, 2)
month ← edate.substring(3, 5)
year ← edate.substring(6, 10)
adate ← month + "/" + day + "/" + year

FIGURE 8.31   Converting a European date to an American date.

would be written as 04/13/1965 by an American and as 13/04/1965 by a European. Consider writing a website that requires users to enter a date. Perhaps we require the user to enter their birthdate or the date that their driver's license was granted. We might want to allow a European to enter a date using the European format but then convert the date to an American format so that it can be stored in our server's database using the same format as American users. We can use text-processing operations to perform this conversion.

We create our web page with a text field that allows users to enter text that we assume to be a date. Our web application code must take any text entered by a European user and convert the date to the American format. This task consists of several specific steps that include (1) extracting the month, (2) extracting the day, (3) extracting the year, and finally (4) constructing the American date. These steps are given by the code of Figure 8.31.

The first step is to obtain the text that the user has entered into the text field. This action is denoted by the readDateFromUser function. Once the date string has been obtained we extract the day, month, and year substrings. The final statement uses string concatenation to construct a new string that follows the American date format, the month preceding the day.

8.4 PATTERNS

Patterns are a very useful technique for processing textual data. A pattern defines a set of properties that some strings will possess and other strings will not. In other words, a pattern is a way of determining whether a particular string is a member of the family defined by the pattern or whether a particular string is not a member of the family.

Figure 8.32 illustrates the concept of a pattern. The string literals in the first column are easily recognized as following a well-known pattern.

Common Textual Patterns	
String Literal	**Pattern Family**
"123-45-6789"	Social Security number
"999-999-9999"	Phone number
"beatles@sub.edu"	e-Mail address
"04/13/1963"	Date
KMOX	Radio station call sign
"1234-1234-1234-1234"	Credit card number
"XOXO"	Hugs and kisses
PG-13	MPAA movie rating

FIGURE 8.32   Common textual patterns.

The string literal "123-45-6789", for example, follows the pattern that all Social Security numbers must follow. The pattern of Social Security numbers is understood to be any 3 digits followed by a dash (-) followed by any 2 digits followed by a dash followed by any 4 digits. A string literal that matches this pattern can be reasonably understood as a member of the Social Security number family, whereas a string literal that does not match this pattern is not a Social Security number.

In computer programming, patterns are known as regular expressions. A *regular expression* defines a pattern such that a particular string will either match the pattern or will not match the pattern. Regular expressions are extremely powerful techniques for processing textual data; particularly for finding elements in large textual databases. We will first describe how to construct a pattern and we will then show how a pattern can be used to find small pieces of text that are hidden in very large textual databases.

8.4.1 How to Write a Pattern

The following rules describe how we can write a valid pattern and how to know the meaning of any pattern that we write (Figure 8.33). We will expand this list of rules in later subsections so that we will be able to write more concise and more powerful patterns.

8.4.1.1 Case Study: Hugs and Kisses Pattern

Consider writing a pattern for the hugs and kisses string as described in Figure 8.32. The hugs and kisses pattern is the simplest kind of pattern we can write since the pattern is so narrow that there is only one string literal

Basic Rules for Writing a Pattern

1) Any single character (except for a few *special* characters) is a pattern. The character represents itself.

2) The period symbol (.) is a pattern. The character represents any single character. The period is special.

3) If A and B are both patterns, then so are

   a) AB : Sequencing represents the pattern A followed by the pattern B

   b) A|B : Alternation represents either the pattern A or the pattern B

   c) (A) : Grouping sets aside pattern A as a group. The parentheses are *special*.

FIGURE 8.33   Basic rules for pattern writing.

that matches the pattern. We refer to this as a fixed pattern, a pattern with no variability.

We define the pattern as XOXO. We can prove that this is a pattern according to the rules of Figure 8.33, by applying the rules to various elements of the pattern. First we note that according to rule 1, the single character X is a pattern and the single character O is a pattern. We then note that XO is a pattern since, according to rule 3a, it is a sequence of the pattern X followed by the pattern O. Finally, we note that XOXO is a pattern, according to rule 3a, since it is a sequence of the pattern XO followed by the pattern XO. *Sequencing* expresses a sequence of subpatterns.

8.4.1.2 Case Study: MPAA Rating Pattern

As another example, consider writing a pattern that describes Motion Picture Association of America's (MPAA) movie ratings system. There are five ratings in this system and they are G, PG, PG-13, R, and NC-17. Using rule 1 from Figure 8.33, we understand that each of the individual characters C, G, N, P, R, -, 1, 3, and 7 are patterns according to rule 1. We can then apply sequencing, given as rule 3a, to form the fixed patterns PG, PG-13, and NC-17. *Alternation* expresses an alternative between two subpatterns and should be read as the word *or*. As a simple example, consider applying rule 3a to the patterns G and PG to obtain G|PG. This pattern should be understood as either a "G or PG". We can apply rule 3a once for every option in our pattern to yield the MPAA rating pattern of G|PG|PG-13|R|NC-17. The meaning of this pattern is that any string that is either a "G" or "PG" or "PG-13" or "R" or "NC-17" is an MPAA rating.

Social Security Pattern

(0|1|2|3|4|5|6|7|8|9) (0|1|2|3|4|5|6|7|8|9) (0|1|2|3|4|5|6|7|8|9)- (0|1|2|3|4|5|6|7|8|9) (0|1|2|3|4|5|6|7|8|9)-
(0|1|2|3|4|5|6|7|8|9) (0|1|2|3|4|5|6|7|8|9) (0|1|2|3|4|5|6|7|8|9) (0|1|2|3|4|5|6|7|8|9)

FIGURE 8.34   Social Security pattern using only the basic rules.

8.4.1.3 Case Study: Social Security Numbers

We will now consider writing a pattern for Social Security numbers. We will see that although it is reasonably easy to describe a Social Security number, the pattern we create will be long and cumbersome. This example serves as motivation for an expanded set of rules that provide more powerful and concise techniques for writing patterns.

As we have already described, a Social Security number can be understood as any three digits followed by a dash followed by any two digits followed by a dash followed by any four digits. There are, of course, exactly 10 digits and we can use alternation to write a pattern that describes a single digit. The pattern is given as (0|1|2|3|4|5|6|7|8|9). This pattern uses grouping to isolate the digit subpattern from other portions of any pattern that contains a digit.

Since we now have a pattern that describes a digit, we can write the social security pattern as shown in Figure 8.34. The pattern literally states something like "a Social Security number is a sequence composed of a digit; a digit; a digit; a dash; a digit; a digit; a dash; a digit; a digit; a digit; and finally a digit." Although the pattern is correct, the pattern is overly long and extremely cumbersome to read. Examples such as this caused computer programmers to develop better techniques for writing patterns. These additional techniques are described in the additional rules that are described in the following subsections.

8.4.2 Repetition Rules

Figure 8.35 gives several additional pattern-writing rules. These additional rules allow programmers to control the number of times an element is allowed to repeat within the context of a pattern.

Repetitions allow us to more concisely rewrite the social security pattern. We recognize that a Social Security number follows the pattern of three digits, two digits, and finally four digits. Since the numbers 3, 2, and 4 denote an exact number of repetitions we may use rule 4d to generate a repeating pattern of digits.

Repetition Rules

4) If A is a pattern then so are

a) A* : Denotes zero or more repetitions of A. The * is *special*.

b) A+ : Denotes one or more repetitions of A. The + is *special*.

c) A? : Denotes zero or one occurrence of A. The ? is *special*.

d) A{m} : Denotes exactly m repetitions of A. The curly brackets are *special*.

e) A{m,n}: Denotes as least m and no more than n repetitions of A.

f) A{m, } : Denotes at least m repetitions of A.

FIGURE 8.35   Repetition rules for pattern writing.

Figure 8.36 shows how we can rewrite the Social Security number pattern using repetition rules. At the beginning of this pattern we use rule 4d to express that a Social Security number must begin with exactly three digits. Following this we express that a dash must be present after which, according to rule 4d, there must be exactly two digits. Finally, the pattern must end with a dash followed by exactly four digits.

Note that a repetition character is applied only to the immediately preceding pattern. In other words, whenever a *, +, ?, {m}, or {m,n} occurs, the repetition is applied only to the immediately preceding pattern. In our example, a digit group occurs immediately before each of the repetition controls and therefore only digits are allowed to repeat. There are no repetition characters that immediately follow a dash, for example, and hence dashes are not allowed to repeat.

As another example that uses repetition, consider expanding our hugs and kisses pattern to include any string that repeats the pattern XO one or more times. We understand that each of the following string literals matches the expanded hugs and kisses pattern: "XO", "XOXO", "XOXOXO", "XOXOXOXO", and an infinite number of others. The pattern (XO)+ describes this expanded hugs and kisses family of string literals since it

Social Security Pattern with Repetitions

(0|1|2|3|4|5|6|7|8|9){3}-(0|1|2|3|4|5|6|7|8|9){2}-(0|1|2|3|4|5|6|7|8|9){4}

FIGURE 8.36   Social Security pattern that makes use of the repetition rules.

denotes one or more repetitions (the plus symbol) of the (XO) group. The grouping parentheses are vital in this example since without them we have the pattern XO+, a pattern that describes any string that begins with a single X and is followed by one or more O's.

8.4.3 Character Class Rules

A *character class* is pattern that concisely defines a set of characters. The term *digit*, for example, names a character class. A digit is defined as the set of the ten characters 0, 1, 2, 3, 4, 5, 6, 7, 8, and 9. Pattern writing rules allow us to define our own character classes and also provide several predefined commonly used character classes that are specially named. These character class rules are shown in Figure 8.37 where each of the subrules gives the special name for a predefined character class.

The predefined character classes of Figure 8.37 merit brief discussion. The class of word characters (\w) contains any character that might reasonably appear in a word. Any of the alphabetic characters, for example, will be in this class but also, in a technical sense, the digits 0 through 9 may also appear in a word. Among the characters that do not belong to the class of *word characters* are symbols such as ?, @, :, #, and many others. These other characters are placed into the nonword character class (\W).

The class of *white space characters* is understood to be characters that are used to separate words and, that when typed into a word processer, generate only white space on the paper. A space character, for example, is a member of this class since when you strike the space bar on a keyboard the

---

Character Class Rules

5) Square brackets denote a character class. Each character within the brackets is a member of the character class. A dash in a character class denotes a range of characters.

   a) \d : Denotes any single digit.

   b) \D : Denotes any character that is not a digit.

   c) \w : Denotes any word character (a-z, A-Z, 0-9).

   d) \W : Denotes any nonword character.

   e) \s : Denotes any white space character (space, tab, line feed).

   f) \S : Denotes any character that is not a white space.

FIGURE 8.37 Character class rules for pattern writing.

Social Security Pattern with Repetitions and Character Classes

\d{3}-\d{2}-\d{4}

FIGURE 8.38   Social Security pattern that makes use of repetition and character class rules.

result is to generate white space in the document you are creating. Other examples of white space characters are tabs and line feeds (given by the enter key on most keyboards).

Character classes give us a very concise technique for writing the social security pattern. First, we recognize that a digit can be expressed, using a character class, as any one of the following patterns:

[0123456789]—This is a pattern that denotes any single one of the characters in the brackets.

[0-9]—This is a pattern that denotes any character in the range 0 through 9 (as denoted by the dash symbol). Note that the dash symbol takes on a special meaning if it is used in the context of a character class.

\d—This is a pattern that denotes any digit. It is equivalent to the class [0-9] and [0123456789].

We can use the digit character class, \d, to denote a single digit within our Social Security pattern. Our final Social Security number pattern is shown in Figure 8.38.

8.4.4 Case Study: DNA Sequencing

Deoxyribonucleic acid (DNA) is a molecule that contains the genetic instructions used by all living organisms. DNA is essential for all known forms of life. Human DNA is composed of billions of discrete pieces of genetic code. Biologists represent the code of a particular DNA molecule as a sequence of the four letters A, C, G, and T. In other words, the DNA code is represented as a string! Each letter in the sequence denotes a particular nucleotide, a base element that serves as a building block for DNA sequences. The letter A represents adenine, C represents cytosine, G represents guanine, and T represents thymine. The length of the DNA string and the order of the letters define the biological function of any part of the DNA molecule.

The Human Genome Project (HGP) is an international research project tasked with determining the DNA sequence of the human genome, among various other subtasks. While generating the general DNA sequence was

Searching a DNA Database

CAGACTTTCAGAACTGTCAGTTCCCCGGATTTTACCCATCACATTTTGCTACTACTTTC
TACTACTATATACTTTTCCAATTTCATACGGGTACTATTATCCATACTCTACTATTAC

FIGURE 8.39   Find the subsequence CATT.

largely completed in 2003, the project continues to analyze the generated data, an effort that is expected to take many more years to complete. The DNA sequence database consists of many gigabytes of data and can be understood, in simplistic terms, as an extremely long string literal.

Research scientists often search through DNA sequences for small subsequences of genetic code. The difficulty of finding a short subsequence of text within a string containing billions of C's, G's, A's, and T's is much worse than finding a needle in a haystack. Figure 8.39 illustrates the difficulty of searching for even a very simple string within a very large DNA sequence. Consider, for example, searching for the subsequence CATT within the small portion of DNA sequence shown in Figure 8.39. The substring CATT occurs exactly once. Can you find it?

Patterns provide a concise technique for searching within text. In the example of Figure 8.39, we can write the pattern CATT and then instruct a computer to identify all occurrences of the pattern in the DNA string. Although a description of how a computer can efficiently find patterns in large textual databases is beyond the scope of this text, computers are indeed able to efficiently search for patterns as long as the pattern follows the rules described throughout this section.

Research scientists often search for substrings that are more complex than fixed patterns such as CATT. A scientist may, for example, be interested in any sequence that begins with AC, is followed by one or more T characters, and finally followed by a C. We can write this pattern as

FIGURE 8.40   Textual matches for the pattern AC.*GAA.*AG.

ACT+C where the subpattern T+ denotes one or more T characters. If we instructed a computing system to find all occurrences of this pattern in the DNA string of Figure 8.39, the computer would produce several results. The substring ACTC occurs once, the substring ACTTTC occurs twice, and the substring ACTTTTC occurs once.

Consider an even more complex pattern such as a sequence that begins with the phrase AC, ends with the phrase AG, and contains the phrase GAA anywhere between the start and end phrases. We can write this pattern as AC.*GAA.*AG. This pattern contains the subpattern period-asterisk. Recall that the period denotes any character and hence the period-asterisk pattern denotes any sequence of zero or more characters. The pattern therefore defines a family of strings that start with the phrase AC, is followed by a sequence of zero or-more characters, is followed by the phrase GAA, is followed by a sequence of zero or more characters, and finally ends with the phrase AG. This pattern occurs in Figure 8.39 where the substring ACTTTCAGAACTGTCAG occurs exactly once.

Figure 8.40 shows how the various elements of the pattern are matched up with the substring that actually occurs in the DNA sequence. Although each of the fixed subpatterns AC, GAA, and AG must only match those specific phrases, the two period-asterisk subpatterns will match any sequence that allows the whole pattern to make sense. In this example, the first period-asterisk matches the sequence TTTCA, while the second period-asterisk matches the sequence CTGTC.

8.4.5 Case Study: Web Searches and Enron Legal Documents

Web search engines such as Google, Bing, and Yahoo allow users to search through massive amounts of textual data by using what amounts to a fixed-pattern search. The web query "popcorn" can be understood as a search for all occurrences of the pattern POPCORN among the text of all documents on the Internet. Search engines of the future will likely support far more sophisticated patterns. Perhaps you would like to find any e-mail address, Social Security number, credit card number, ISBN number, or date that exists on the web. Oddly enough, lawyers and paralegals must often search through massive amounts of textual data looking for sophisticated patterns. We explore the well-known case of Enron, and show how patterns enable sophisticated searches through various legal documents related to the case.

Enron, an extremely large energy company based in Houston, Texas, was formed in 1985 by Kenneth Lay. Several years later, when Jeffrey Skilling was hired, he developed a staff of executives that, through the use of accounting loopholes, special purpose entities, and poor financial reporting, were able to hide billions in debt from failed deals and projects. Chief Financial Officer Andrew Fastow and other executives not only misled Enron's board of directors and audit committee on high-risk accounting practices, but also pressured Andersen to ignore the issues. Shareholders lost nearly $11 billion when Enron's stock price, which hit a high of US$90 per share in mid-2000, plummeted to less than $1 by the end of November 2001. As part of the legal investigation into the scandal, the e-mails of top executives have been made publicly available. We will consider how a legal expert might use patterns to mine the textual database for useful legal information.

People tend to capitalize words they want to emphasize in e-mail messages. The legal team therefore decides to compile a list of words that contain only capital letters. The legal team decides to use the pattern shown in Figure 8.41. The pattern is understood to consist of one or more repetitions of a capital letter. The legal team searches through the massive e-mail database and is surprised by results.

The pattern of Figure 8.41 is incorrect since it does not find whole words that are all caps. This faulty pattern matches any sequence of capital letters without regard to whether the sequence is an entire word or merely a small part of a word. Consider, for example, searching for all occurrences of this pattern in the text of Figure 8.38, a small snippet of data from the actual Enron e-mail database. The pattern does match the obvious all-cap words PRIVILEGED, AND, and CONFIDENTIAL but also matches the phrases S (from Subject), E (from Enron), L (from Livingston), and C (from County). Each of these phrases is a sequence of one-or-more capital letters and hence matches the pattern.

All CAPS Word Pattern (Incorrect)
[A-Z]+

FIGURE 8.41   An incorrect pattern for finding all-caps words.

All CAPS Word Pattern

\W[A-Z]+\W

FIGURE 8.42   A correct pattern for finding all-caps words.

The legal team corrects its pattern by recognizing that a word is bracketed by nonword characters. In other words, the pattern must state that a nonword character must occur immediately before and immediately after an all-caps sequence if the sequence represents an entire word. The legal team recalls that the nonword character class is predefined and denoted as \W. Figure 8.42 shows the corrected all-caps word pattern.

A search for all occurrences of this pattern in the text of Figure 8.43 will now match only the three words PRIVILEGED, AND, and CONFIDENTIAL. The S of Subject does not match the pattern since the character u is not a nonword. Using similar reasoning, we understand that the E, L, and C also fail to match the pattern.

Although the legal team members have created a good pattern for finding any all-caps word, they realize that there are still too many matches in the e-mail database. Words such as I, RE, CC, TO, FROM, and many other similar words occur very frequently and the legal team would like to exclude these words from the results. They finally decide to search the e-mail database for a sequence of all-capitalized words. They determine that the sequence must have at least three words. The legal team therefore generates the pattern shown in Figure 8.44. This pattern

---

Enron e-Mail Database Segment

Subject: Enron - Livingston County PRIVILEGED AND CONFIDENTIAL

FIGURE 8.43   Segment of text from the Enron e-mail database.

Sequence of at least 3 All CAPS Words

\W([A-Z]+\W){3,}

FIGURE 8.44   A correct pattern for finding any sequence of at least three all-caps words.

FIGURE 8.45   Textual matches for the pattern \W([A-Z]+\W){3,}.

defines a sequence of at least three all-cap words. This pattern occurs exactly once in the text of Figure 8.42. The phrase " PRIVILEGED AND CONFIDENTIAL" matches the pattern.

Figure 8.45 shows how the phrase "PRIVILEGED AND CONFIDENTIAL" is a match for the pattern of Figure 8.44. The first character is a space and matches the first \W of the pattern. Following the space are three occurrences of the [A-Z]+\W pattern. Each of these three elements begins with a sequence of at least one capitalized letters and is followed by a space.

REFERENCE

1. von Neumann, John. "First Draft of a Report on the EDVAC," 1945.

TERMINOLOGY

alternation	min
arity	namespace
associativity	operand
average	operator
cell	pattern
character class	precedence
data dependency	range
fixed program architecture	regular expression
formula	repetition
function	right-associative
left-associative	self-modifying code
max	sequencing

spreadsheet

stored program architecture

string

  concatenation

  indexing

  indexOf

  length

  literal

substring

sum

value space

von Neumann architecture

white space characters

word characters

worksheet

EXERCISES

1. Give the value produced by each of the following formulas. In addition, show the result of every discrete step you take in the evaluation process by constructing a figure similar to Figure 8.11. Make sure that you follow the precedence and associativity rules described in this text.

  a. $3 + 2 \wedge 3 * 4$

  b. $5 * 3 * (2 + 3) - 4$

  c. $1 + 1 + 4 / 1 + 1$

  d. $3 * 3 + 3 - 3 / 3$

  e. $5 * (3 + (6 / 3 - 1))$

2. Give the result of the expression "1 − 0 − 1" if we assume that

  a. Subtraction is a right-associative operator

  b. Subtraction is a left-associative operator

3. Answer each of the following questions with respect to the spreadsheet shown in the following figure.

3.

	A	B	C	D	E
1	5	3	3	=sum(A1:C1)	=D1/3
2	2	6	4	=sum(A2:C2)	=D2/3
3	1	5	3	=sum(A3:C3)	=D3/3
4					

    a. What values do the formulas in cells D1, D2, D3, E1, E2, and E3 display?

    b. What formula should be entered into cell D4 to compute the sum of the values in column D rows 1 through 3? Make sure to use a function.

    c. What formula should be entered into cell E4 to compute the average of the values in column E rows 1 through 3?

    d. Draw the dependency diagram for the formula of D4 that you constructed in part (c).

    e. Write a formula to compute the maximum of the numbers in columns A through C, rows 1 through 3. Make sure that your formula will work even if any (or all) of the values in that range change at some later time.

    f. Describe the nature of the error that would be created if a user entered the formula "=sum(A1:D1)" into cell A1.

4. Consider using indexing, length, indexOf, substring, and concatenation to perform the following tasks. For each task, describe a step-by-step procedure using these string processing functions to accomplish the task.

    a. Consider a string of your three-letter initials. Write a procedure that takes any three-letter initial string and creates a two-letter initial string containing only the initials of the person's first and last names. The three-letter initial string for Martin Luther King is given as "MLK", for example. Your procedure should generate the string "MK" when given these initials. Your procedure must work for any three-letter initial string.

    b. Convert a string that contains a person's first name, followed by a comma, followed by the person's last name, and reverse the order of the names. Your procedure must work regardless of what string is processed by your code.

    c. The first part of a URL contains what is known as the *scheme*. Commons schemes include http, https, and mailto. A colon always occurs immediately after the scheme in order to separate the scheme from the remaining parts of the URL. Write a procedure to extract the scheme for any URL.

d. Credit card numbers typically range in length from 14 to 19 digits depending on the issuer. The first 6 digits are known as the Issuer Identification Number (IIN) and the remaining digits constitute the individual card holder's account number. Write a procedure to extract both the IIN and the card holder's account number from any credit card string. You may assume that the string contains only the digits of the card without any hyphens or other separators.

e. Tic-tac-toe is a game where users take turns writing either an X or an O on a 3-by-3 grid. The objective is to get three marks in a row. Consider representing the state of a tic-tac-toe game as a string of nine letters where each letter is an X, an O, or a space. The nine letters are arranged row by row. Write a procedure to create the two three-letter diagonals for any tic-tac-toe string. The string "XO OXOXOX", for example, has the two diagonals "XXX" and " XX".

5. Write a pattern for each of the following string families. Make sure that your pattern matches a string only if it is a member of the family and that it does not reject strings that are members of the family.

a. Any binary number. A binary number is a string of any length that contains only the digits 0 and 1. A binary number cannot start with a 0, however, unless it is the only digit in the number. Examples of valid binary numbers include:

 - 0

 - 1010

 - 101101000101010101010101010101000001010111111

b. A phone number. For this exercise, a phone number has a three-digit area code that is optionally enclosed in parentheses. After this comes a space followed by three digits followed by a dash followed by four digits. Examples of valid phone numbers include:

 - (555) 329-1559

 - 123 456-7890

c. The call letters for an American radio station. American radio stations are known by either a three- or four-letter call sign. The first letter of the call sign is a K for stations west of the Mississippi river and a W for stations east of the Mississippi. The remaining letters of the call sign are all capitalized. Examples of valid radio station call letters include:

- WXHO

- KDKA

- KMOX

- WWV

d. United States pay grade classification. Personnel employed by the United States government are assigned a pay grade that determines their annual salary. The pay grade is a code beginning with the letter E, W, or O. Following the first letter is a dash followed by either one or two digits. Valid E pay grades move from 1 up to 9. Valid W pay grades move from 1 up through 5. Valid O pay grades move from 1 up to 10. Examples of valid pay grade classifications include:

- E-1

- E-9

- W-5

- O-1

- O-10

e. Any Roman number ranging from 1 up to 10. Examples of valid Roman numerals in this range include:

- I

- II

- IV

- X

6. Consider a string containing only the digits 0 and 1. For each of the following items, write a pattern that matches.

   a. All strings that begin and end with a different digit. Examples include:

      - 01

      - 0111010001

      - 1000101010101010

   b. All strings that contain only even-length runs of zeros. Examples include:

      - 1001100001

      - 00001100

      - 000000

7. Write a pattern for each of the following items.

   a. All words that start with *m* and end with *ing* and that are exactly seven letters long.

   b. All words that start with *b* and do not end with *e*, *d*, or *k* and are exactly four letters long.

   c. All words that consist of only the letters *a*, *e*, *i*, *o*, *n*, and *t* and are at least three letters long.

   d. Any word that contains the letter *n* exactly three times.

8. Simplify each of the following patterns.

   a. (X+|X)O{1}X

   b. (staff|drill|staffsargent|drillsargent|staffmaster|drillmaster)

   c. (O|X|X{1,}|O{1,}){0,}XXXXXXXXX+(O|X|OX|XO)+<*Problems U1466O3197E End Here*>

CHAPTER 9

Let's Get It Correct

It is much easier to be critical than to be correct.

—BENJAMIN DISRAELI

OBJECTIVES

- To know that many, if not most, of so-called computer errors are really human errors, generally data entry errors
- To know that correctness is impossible without a specification defining what correctness means
- To be able to explain the difference between verification and validation, and the reasons that each is important
- To realize that it is possible to prove the correctness of some software, but that such proofs are only used in rare occasions due to their cost and complexity
- To know the limitations of software testing with respect to correctness
- To be able to diagram a basic test suite consisting of test cases defined in terms of input conditions and expected behavior
- To be able to explain the difference between white box and black box testing, and the advantages of each
- To be able to analyze a test suite for a given activity diagram and demonstrate statement coverage
- To be able to explain the difference between statement coverage and path coverage and why path coverage is typically impossible
- To recognize equivalence partitioning and its role in testing
- To be able to construct a test suite using boundary value testing for one or two simple inputs

One vendor selling various computer software packages advertises the following: "Fix Your Computer Errors in Minutes!" But just what does this company, or anyone else trying to assign blame to a defenseless machine, mean by the term *computer error*. This chapter is all about trying to identify the various problems that people label as computer errors. Further, we shall look at things you should know about how to minimize such problems.

9.1 "COMPUTER ERRORS" USUALLY AREN'T

It is an unfortunate fact that humans regularly describe things as computer errors or glitches that are in fact human errors. Perhaps it is because we want to avoid responsibility for problems, or perhaps it stems from ignorance, even a kind of fear of unknown inner workings of computers, or perhaps it is simply the appeal of allowing a machine without the ability to defend itself to serve as scapegoat. Whatever the reason, computers seem to be a frequent target. So at the beginning, it is crucial to remember that the likelihood of human error is considerably more common than computer error. If the bank makes an error in your checking account, the probability of a computer calculation being in error is extremely remote by comparison to the probability that a bank employee entered incorrect information.

Figure 9.1 suggests seven reasons for what are commonly called computer errors. Of the seven only two—hardware failure and software fault—can fairly be blamed on computers, and these two are probably the least likely. The first three, and by far the most common, reasons are best understood as human error. *Incorrect data entry* occurs anytime that someone provides incorrect input to a computing device. When a bank teller enters the wrong amount for your deposit; when you are selecting an option from a pull-down menu and accidentally choose the state Wyoming when you meant to pick Wisconsin; when the physician erroneously enters your sister's name while trying to access your medical records—all of these

- **incorrect data entry**
- **system misunderstanding**
- **improper interpretation of computer output**
- **physical damage**
- **hardware failure**
- **software fault**
- **security breach**

FIGURE 9.1   Reasons for "computer errors."

situations are examples of incorrect data entry. Of all reasons for computer error, data entry error is undoubtedly the most common. The simple truth is that humans make errors, and the more data we are expected to input the more likely we are to make errors.

Sometimes human error can be explained as *system misunderstanding*, because a system user or administrator fails to properly comprehend how the system functions. Here are a few common examples:

- An e-mail user does not understand the difference between the Reply and Reply All buttons.

- A browser autocompletes a word's spelling differently than intended.

- An autofocus camera produces blurred photos because the user does not understand that freezing motion requires higher shutter speeds.

- The small keyboard of a smartphone results in typing errors.

- The computer misinterprets track pad input due to an accidental touch of a second finger.

- A store clerk politely apologizes for the need to reenter your purchase a second time due to the use of a new checkout system.

We have all found ourselves confused from time to time by a new computer application and a typical response is to guess. If we guess wrong, as is frequently the case, the result is an error. These kinds of errors could sometimes be called data entry errors, but they are distinguished here because they result from ignorance—a factor that can often be addressed.

The third reason for computer errors that is certainly a human error is *improper interpretation of computer output*. Just as you can type something wrong for input, you can also misread computer output. Perhaps you overlooked an extra zero in the price of that online purchase. Perhaps an income tax preparer read the wrong row from the tax table display. Perhaps road noise causes a driver to misinterpret the audible directions from a GPS unit.

The processing power of computers makes possible extremely complex calculations, but interpreting all of that computer-created results can be a daunting task. We can examine Doppler radar data and satellite cloud cover images, but predicting weather is still full of inaccuracy. Computers can analyze stock market history in limitless ways, but tomorrow's Dow Jones numbers are still a mystery.

Although *physical damage* to your computer is not exactly human error, it still often results from human negligence. There are three common types of physical damage: jarring/crushing, water, and fire.

Many of the components of a modern computer are sensitive to jarring motions. Most of us know people who have broken LCD screens by dropping laptops or cell phones. Disk drives, and even some electrical circuitry, can also be damaged by jarring or crushing.

Most electrical devices are susceptible to water damage, especially if the device is powered on while being soaked. The best advice for a wet computer is to turn the device off, open the case to expose as much of the circuitry as possible, and leave it in a dry place for several hours before attempting to power on.

Fire damage depends upon temperature. Computers contain many plastic, composite, and metal parts—all of which have melting temperatures achieved by common building structure fires.

By *hardware failure* we refer to errors that are caused by the computer hardware but not resulting from aforementioned physical damage. The truth is that modern electronics rarely fail, and if they do, such failure generally occurs within the first few weeks. Most nonmechanical components do not wear out with the exception of batteries that tend to degrade over time. In addition, error checking mechanisms within computer memories further reduce the likelihood of memory storage errors.

So what kind of hardware errors are possible? The most vulnerable parts of a computer are those that are mechanical: disk drives, CD/DVD players, and mice. It is not uncommon for a disk drive to fail after several years of service, resulting in the potential loss of the data it stores.

There are other rare events that fall under the heading of hardware error. Power outages obviously shut down reliant electrical systems. Even spurious gamma radiation or powerful magnetic fluctuations have been known to disrupt certain parts of a computer under extremely rare circumstances. Most of us are unaffected by these unlikely events, but the interconnection of millions of computers via the Internet increases the possibility that you might encounter the occasional server hardware error.

Of the two reasons that are most fairly categorized as computer errors, *software faults* are more common than hardware failure. A software fault is sometimes called a *bug* or *flaw*, and is the result of an incorrect computer program. Most of the remainder of this chapter expands upon the concept of software faults.

The final reason mentioned that leads to computer errors is *security breaches*. A security breach occurs whenever a computer is successfully attacked and caused to perform in unintended ways. Chapter 12 of this book is devoted to computer security, including an extensive discussion of how security breaches occur.

9.2 SOFTWARE CORRECTNESS

When people refer to computer "correctness," they usually mean software correctness. As mentioned in Section 9.1, hardware errors are relatively rare. So if it is the computer (not human) error that is of concern, then software is generally the culprit. The truth is that virtually every computer program of significant complexity contains flaws. Fortunately, most of us do not encounter these flaws, so for the most part computer systems behave correctly.

Before delving further into the issue, we must clarify the meaning of the word *correctness*. In and of themselves computers are just machines that perform (via software) what they are "told." This means that a computer system or program is neither correct nor incorrect by itself. The user may not like how the program behaves, but that does not necessarily mean the program is incorrect. Suppose that your word processor automatically capitalizes words at the beginning of a new sentence. You are typing a list of terms, one per line, and you do not want the word processor to capitalize them all. Clearly, you find the autocapitalization feature annoying, but does the annoyance mean the program is incorrect? Consider your friend who types short stories and loves the autocapitalization. To your friend autocapitalization is both correct and appreciated.

The point is that computing systems and computer software can only be considered to be correct or incorrect in the context of some particular intent. A system is correct if its behavior matches the intended behavior. In other words, it does what it is supposed to do.

When a computer program is created, the person(s) for whom it is created is known as the *customer*. Properly engineered software is designed to meet the customer's wishes. As explained in Chapter 4, one of the first tasks in the process of engineering software is to define the problem. The resulting set of requirements is intended to represent the wishes of the customer. Unfortunately, written requirements are not always a good match with customer desires. Maybe the customer changed his or her mind. Maybe time has passed and the goals for the software have changed.

As is often the case, maybe the requirements just were not properly ana-lyzed. Whatever the reason, it is all too common for customer wishes to be somewhat different from written requirements. This leads to two different kinds of correctness:

1. Software can be correct with respect to the customer's wishes.

2. Software can be correct with respect to the written specifications. (We define *specifications* as some documented form of intended soft-ware behavior.)

Software engineers refer to the process of ensuring correctness as *veri-fication and validation (V&V)*. Validation ensures that the product meets the needs of the customer—the first kind of correctness listed earlier. Verification ensures that the product meets its specifications—the second type of correctness. Figure 9.2 illustrates V&V.

Just as it is easier to understand a legal document than it is to understand its *intent*, so too it is easier to verify software systems than it is to validate them. Verification is possible once we have a written set of specifications. Software requirements are a common form of specification, but they are not the only kinds of specification. As we saw in Chapter 6, computer sci-entists can specify the behavior of computer programs in many ways, using pictures, logic, and natural language. Any of these documented forms of intended behavior could potentially be used in verifying correctness. Most of the remainder of this chapter is devoted to such verification tools and techniques. That said, there are still a few useful ideas regarding validation.

Customer involvement in the software development process is perhaps the most critical factor in achieving validation. Some software engineering

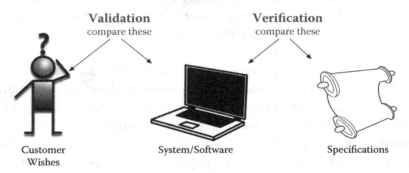

FIGURE 9.2   Two kinds of correctness and how they are ensured.

processes have found that it is best to include the customer (or customer's representative) as a member of the software development team. This means that if a bank contracts with a software firm to develop a new online checking system, then it might be wise for the bank to assign one of its knowledgeable employees to the project. Similarly, an automobile manufacturer should probably include one of its engineers to the outsourcing company that develops new traction control software. This integral involvement in software development is yet another reason why noncomputer scientists need to understand computer science.

Some other techniques that might involve you as part of a validation process are as follows:

- *Beta testing*—Enlisting selected users to use and comment on a nearly complete system

- *Usability testing*—Employing individuals to explore the look and feel of a system

- *Acceptance test*—Any process performed by customers to validate a system before final purchase

Customers should expect a reputable software development team to be responsible for some kind of verification, but responsibility for validation tends to fall to the customer. Beta testing, usability testing, and acceptance testing are the current best practices for validation. All three of these rely upon involvement of people who are not computer scientists and often whose primary employment lies outside the company developing the software. Many of the techniques used for verification, and described in following sections, can also be applied in these validation tests.

9.3 VERIFICATION

Ultimately, it is customer wishes that truly define correctness. However, customers often do not know precisely what they want or are unable to clearly express their desires. To make matters worse, the needs of a customer can change as time passes. All of these factors point to the need for well-written requirements. The requirements are intended to be a physical representation of the customer's wishes. Because of documented requirements, a customer can be confident of what is being built. Because of documented requirements, software developers know what to build.

As explained in Chapter 4, requirements need to be clear, consistent, and complete. Otherwise, they are impossible to verify. A good way to determine the acceptability of a requirement is to ask the question, "Is this requirement verifiable?" To be verifiable means that there is a sound means for ensuring the software meets the requirement.

Imprecision is the primary reason for poor, unverifiable requirements. For example, consider the following requirement:

Requirement S1—The program must be secure.

S1 is simply too vague to be verifiable. What did the author mean by the word *secure*? Is this type of security concerned with privacy, or immunity from system failure, or some other kind of security? The following two requirements are both more precise to the point of being verifiable:

Requirement S2—All users must be required to log in with a user name and password.

Requirement S3—All user-related data stored by this program will be encrypted.

As a second example, consider the following vague requirement relating to system performance:

Requirement E1—The program must be fast.

Although E1 is clearly unacceptable, there are ways to express performance so as to make it verifiable. Here is an example of an acceptable alternative requirement:

Requirement E2—The system will respond to all user input with a new query within .5 seconds.

The unacceptable requirements S1 and E1 are both examples of *nonfunctional requirements*. Nonfunctional requirements, as previously described, are used to express things like performance, security, portability, reliability, safety, documentation, delivery, and user-friendliness. By comparison, functional requirements express specific behaviors of the software. You can think of it this way: functional requirements define *what* the software does, whereas nonfunctional requirements define *how well* it does so. The nonfunctional requirements tend to be more difficult to define in verifiable ways, which means that good software engineering must pay careful attention to both functional and nonfunctional requirements.

Once precise requirements, or any other precise specifications, have been formulated, there are two ways that software engineers verify correctness:

1. Proofs of correctness

2. Software testing

A *proof of correctness* is the only way to guarantee that software is correct with respect to its specifications. The process of proving a program to be correct is similar to proving a mathematical theorem. A computer scientist uses a precise set of rules and manipulations to argue that when the program executes, it must behave exactly as specified. The rules that are used are based upon the known behavior of each individual programming language's instruction and upon mathematically accepted practices in deductive and inductive reasoning (logic).

In practice, programs are rarely proven to be correct. The reason for this is largely cost. For any nontrivial program a proof of correctness requires considerable computer scientist time. Furthermore, proofs are impossible unless the specifications are stated in the most formal logic, and expressing specifications in this way is also time consuming. Another downside is that only the most gifted computer scientists can prove software correctness reliably and even they can make errors in their proofs.

A final problem with proofs of program correctness is the very nature of an algorithm. Propositional logic is built upon propositions (facts). However, in traditional propositional logic there is no accounting for execution, that is, the passage of time that changes the facts. Computer programs, on the other hand, are written for the purpose of changing state with the execution of each instruction. This notion of time passage, together with some of the more complex data structures, makes it more challenging to formally reason about computer programs.

Even with all of the drawbacks, proving a program to be correct is the only way to be absolutely certain that the software satisfies its specifications. There are some instances when the high cost is justified, usually because of the costs associated with software failure and typically when the highly sensitive code is relatively short. Transportation guidance systems and highly sensitive medical instrumentation are examples of situations where software errors can be life threatening and, therefore, might justify proving parts of the software to be correct.

9.4 SOFTWARE TESTING

If we cannot afford to prove that a program is correct, then we resort to the next best alterative: *software testing*. As the name implies, software testing consists of executing a program and observing its behavior. The exhibited behavior is then compared to the behavior that is required according to the specifications. A successful test shows that the software behaves according to the specifications, and an unsuccessful test (software fault) is demonstrated by a test that discovers a behavior inconsistent with the specifications.

The most important thing to remember regarding software testing is that for any nontrivial program *software testing can never guarantee software correctness*. Even when testing is carefully applied, there are plenty of examples of software failure. On February 11, 2007, twelve F-22 Raptor aircraft (also known as stealth fighters) were deployed for the first time to Asia. Upon crossing the international date line all 12 fighters' systems shut down, including navigation, some communications, and even the fuel systems. This failure resulted from a software bug. Apparently, the software had never been tested for crossing the date line. Fortunately, the aircraft were escorted by tanker planes that were able to return them safely to Hawaii. This is but one of countless examples of software faults not captured during testing.

The reason that software testing cannot guarantee correctness stems from the fact that each test case consists of a single program execution under a single set of circumstances. If the test case succeeds, then the tester can only conclude that the program works for this one situation; the program might still fail under different circumstances.

Think, for example, about a program to log in users, using a typical login panel like the one shown in Figure 9.3. We could construct a *test case*

FIGURE 9.3   A login panel.

Input Conditions	Expected Behavior
User Name = Alisha Armstrong Password = 2Dogs@Home	The program should display a *valid login* message.

FIGURE 9.4   A test case for the login panel.

using a valid user name and associated password. Figure 9.4 shows one way to express such a test case. The input conditions of this test case are variable values or, in this case, user input that defines the test case, and the behavior column explains how a correct program should behave for the given input conditions. In this case we can presume that a user name of *Alisha Armstrong* paired with a password of *2Dogs@Home* constitutes a valid login combination.

If a test case such as that from Figure 9.4 is successful (i.e., if the program executes for the given input conditions and produces the prescribed behavior), then the tester knows that the program worked for this one situation. However, the program might still not work for a different valid user name and password, or the program might erroneously allow some illegal name and password combination to log in. The only way to be completely certain of correctness would be to test all possible user name and password combinations, but such a large number of test cases is impossible to administer in a reasonable period of time.

So if one test case only considers one situation, then how is it possible to use test cases to gain confidence in correctness? The answer to this question is analogous to the concept of polling. Pollsters know that predicting the outcome of a political election cannot be done using the opinion of a single voter. However, polling several carefully selected voters can provide confidence in the outcome prediction. In the same way software testers must use not one but several test cases, and these test cases should be carefully chosen in ways that provide confidence in the correctness of the software. Such a group of test cases is known as a *test suite*. Of course, no test suite will absolutely ensure correctness, which is why software failures still occur. However, effective test suites will capture many software errors. Later sections of this chapter examine ways to design test suites to make them more effective.

Not only are test cases grouped into suites, but testing should also be done at each stage of software development. This comes from another special form of divide and conquer called *layering*. A layering approach to problem solving is like peeling an onion. Inside the onion are many layers and each can be peeled off separately. One way to think of software development layers is to consider the individual components written by programmers as one layer. A group of components form a software module at another layer. The modules form subsystems at a third layer, and the subsystems form the complete software system at a final layer. A layered approach to testing suggests that the software should be tested at each layer. This means that individual programmer's components should be thoroughly tested before they are combined into components for a component-level test. Similarly, modules and subsystems are tested by separate test suites, before a final system test suite is applied. Usability, beta tests, and acceptance tests provide yet more layers to testing. Each layer of testing can focus on different aspects of the software, leading to improved confidence in correctness. A side benefit of the layered approach to testing is that some layers can be performed early in the development cycle, which allows errors to be corrected with lesser cost.

Creating good test suites can be as difficult as developing the code. Therefore, good test suites are carefully maintained so that they can be reapplied in future versions of the software. This reapplication of test suites is so common that it is called *regression testing* by computer scientists. As computer programs are altered to add new features or correct bugs, it is essential that regression tests be executed to avoid the introduction of new bugs.

Another of the challenges in software engineering is that software testing is by nature contradictory to software development. Software development is intended to build a product that works, whereas software tests are designed to discover situations in which the product does not work. Computer scientists are trained to engineering correct programs, but the job of a tester is to find test cases demonstrating incorrectness.

This dichotomy of software development and software testing means that it is often best to separate testing from other software engineering. As much as possible, it is best to use separate teams for testing software. Sometimes this means that testers from outside companies should be used; sometimes the testers are even noncomputer scientists. Larger software development firms establish in-house testing, often performed by so-called *software assurance* teams.

All software testing can be partitioned into one of two categories: black box testing and white box testing. *Black box testing* means that the tester has no access to the actual program instructions. The software is a black box in the sense that testing proceeds by executing the code without any additional information about the program's inner workings. The absence of this knowledge causes the tester to rely primarily upon functional requirements when crafting test cases; so black box testing is also known as *functional testing*. In contrast to black box testing, *white box testing* requires knowledge of the instructions that make up the software. White box testing is examined in more detail in Section 9.5; Section 9.6 is devoted to black box testing.

9.5 WHITE BOX TESTING

The effectiveness of testing depends upon the particular selection of test cases that make up the test suite. One way to select a good collection of test cases is to examine the structure of the program, in particular the control flow of the program is used for white box testing, also known as *structure testing*.

The very minimum that could be expected from a white box test suite, called *statement coverage*, requires that every one of the program's instructions be executed by at least one of the cases in the test suite. In other words, the test suite cannot allow some part of the code to go unexercised. Remember, however, that just because an instruction has been executed successfully once does not mean that the same instruction might not fail under different circumstances.

A more effective form of white box testing is called *path testing*. This form of testing requires an analysis of control flow and can be used to evaluate any process that is described by an activity diagram. Figure 9.5 is an activity diagram for completing an online purchase repeated from Chapter 6.

Complete path testing dictates that every activity diagram path from start to finish be executed by at least one test case. For the activity diagram of Figure 9.5 complete path testing necessitates four test cases: one for PayPal and one for each of the three different credit card companies.

Another example is illustrated by the activity diagram of Figure 9.6, also repeated from Chapter 6. Here there are three paths through the activity diagram corresponding to the three possible driving routes. One path results from a good traffic report for the Baltimore-Washington Pkwy. The other two paths correspond to a bad report for the Baltimore-Washington Pkwy: one for US-50 and one for US-29. Figure 9.7 shows the test suite of three test cases for such complete path coverage.

To assist in examining path testing it is helpful to add a unique number to every activity, as shown in Figure 9.6. The test suite from Figure 9.7 can

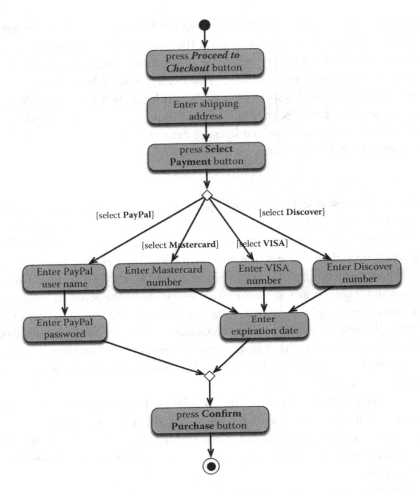

FIGURE 9.5 Activity diagram for checkout of an online purchase.

be analyzed case by case according to the numbered activities that are executed each the path:

Test Case	Executed Activities
Baltimore Pkwy Test Case	1-8
US-50 path	1, 9-15, and 6-8
US-29 path	1, 9-13, and 16-19

Such numbering of activities/instructions can also be used to justify that statement coverage has been achieved by a test suite. Figure 9.8 shows a table that contains a check for every activity that is executed by the test

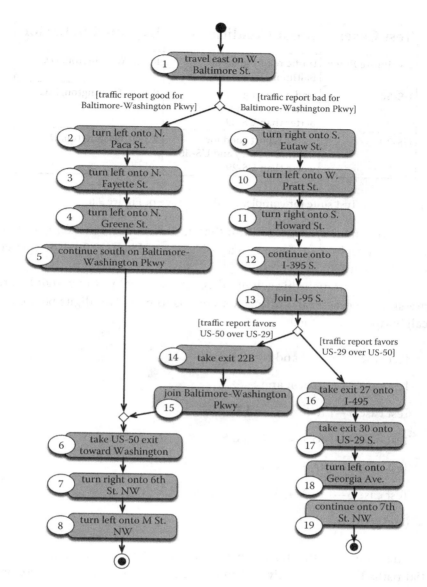

FIGURE 9.6 Activity diagram for traveling from Baltimore, Maryland, to Washington, D.C.

case for that row. In the event that all columns contain at least one check, as in this case, statement coverage is accomplished.

The idea behind complete path testing is that every possible combination of instructions has been attempted, but complete path testing is not always possible. In particular it is impossible to achieve complete path

Test Cases	Input Conditions	Expected Behavior
Baltimore Pkwy	traffic report good for Baltimore Pkwy	Reach Washington, D.C.
US-50	traffic report good for Baltimore Pkwy and US-50 better than US-29	Reach Washington, D.C.
US-29	traffic report good for Baltimore Pkwy and US-50 worse than US-29	Reach Washington, D.C.

FIGURE 9.7   Test suite for complete path coverage of Figure 9.6.

testing on virtually any algorithm that includes loops. For example, consider the telephone dialing activity diagram of Figure 9.9 (also reprised from Chapter 6).

One path through this activity diagram is to dial just one digit before pressing the green call button. A second is to press two digits before the call button, and so forth:

Test case 1—1 to 3, and 5

Test case 2—1 to 3, 4, and 5

Test case 3—1 to 3, 4, 4, and 5

Test case 4—1 to 3, 4, 4, 4, and 5

Test case 5—1 to 3, 4, 4, 4, 4, and 5

Test case 6—1 to 3, 4, 4, 4, 4, 4, and 5

(and so forth)

The problem is that for any particular path there are additional potential paths that repeat activity 4 more times. We can conclude, therefore,

Test Case	1	2	3	4	5	6	7	8	9	10	11	12	13	14	15	16	17	18	19
Paypal	✓	✓	✓	✓	✓	✓	✓	✓											
US-50	✓				✓	✓	✓	✓	✓	✓	✓	✓	✓	✓	✓				
US-29	✓								✓	✓	✓	✓	✓			✓	✓	✓	✓

FIGURE 9.8   Ensuring statement coverage for Figure 9.6.

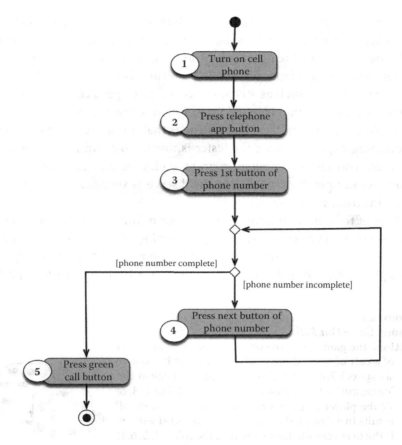

FIGURE 9.9 Activity diagram for dialing a phone number.

that there are an infinite number of paths, and that makes complete path testing unachievable. Since most software incorporates loops, programmers do their best to approximate path coverage by at least achieving statement coverage and also including more test cases to explore multiple loop body repetitions.

9.6 BLACK BOX TESTING WITH EQUIVALENCE PARTITIONING

If you cannot examine the inner structure of the program, either because it is not available or because you are not a computer scientist, a second kind of testing—black box testing—is still an option. Performing black box testing is no different than white box testing in the sense that testers execute each case of a test suite and compare actual behavior to the intended behavior. The difference is how the test cases are selected. Since testers are unable to examine code structure, black box testing cannot utilize techniques such as statement coverage or path coverage. Instead, all test case generation is based upon software requirements.

A big advantage of black box testing is that it requires little or no programming expertise, since the tester is unable to examine the code. For example, consider a computer program to play the dice game called craps. For this example, the requirements will serve as specification. Figure 9.10 contains these requirements.

One of the common techniques for determining test cases begins with a concept known as *equivalence partitioning*. The idea of equivalence partitioning is to separate all possible input conditions into separate groups in such a way that each group has a commonality with respect to

Index: *C1*
Name: *Come Out Roll*
Action: The game of craps consists of one or more rolls of a pair of 6-sided die (each side of a die has one to six spots.) The first roll of a new game is called the "come out" roll. If this roll results in a total of 2, 3, or 12 the player loses by **crapping out**. If the first roll results in a 7 or 11, then the user is a **natural** winner. If the user rolls anything else (i.e., the roll is 4, 5, 6, 8, 9, or 10), then this dice total becomes the point and the next roll is a **point roll** (see Requirement C2).

Index: *C2*
Name: *Point Roll*
Action: The user rolls the dice again. If this roll results in a total of 7, then the user loses by **sevening out**. If the roll total is the same as the point (from the come out roll), then the user wins by **making the point**. If the roll totals anything but 7 or the point, then the point roll is repeated.

Index: *C3*
Name: *Continued Play*
Action: Following a crap out, natural winner, seven out or making the point, the user may begin a new game. (See Requirement C1 for the first roll of a new game.)

FIGURE 9.10   Requirements for a craps game.

the specifications. For the come out roll of the craps game there are four different groups of input conditions that seem best:

1. Crapping out group—Input condition is a roll of 2, 3, or 12

2. Natural winner group—Input condition is a roll of 7 or 11

3. Point group—Input condition is a roll of 4, 5, 6, 8, 9, or 10

4. Error group—Input condition is a roll less than 2 or greater than 12

Notice that rolls of 2, 3, and 12 are placed in the same group, because they all produce the same result (a crap out). Similarly, 7 and 11 produce a win for the player and other valid rolls establish a point, requiring additional dice rolls. Thorough testing must not overlook invalid situations. The error group is included for this purpose. A dice roll less than 2 or greater than 12 is impossible with physical dice, but in a computer program such things might happen.

Equivalence partitioning is a bit like cutting a pizza. You would prefer that every pepperoni on the pizza end up in one and only one slice. In the same way every possible input condition must end up in one and only one group of an equivalence partitioning (Figure 9.11).

Building a test suite from equivalence partitions requires that you select at least one test case for every group. Figure 9.12 shows a test suite for the come out roll that has been chosen using equivalence partitioning.

Notice that these test cases are not exhaustive. A roll of 5, for example, is never considered in the four test cases from Figure 9.12. Also, there is nothing special about which particular value is selected from an equivalence group. For example, the crapping out test case could just

FIGURE 9.11   Slicing pizza is like equivalence partitioning.

Test Cases	Input Conditions	Expected Behavior
Crapping Out	coming out roll is 3	the user craps out
Natural Winner	coming out roll is 11	the user is a natural winner
Point	coming out roll is 4	the user establishes a point and is required to roll again
Error	coming out roll is −4	the program reports an invalid roll

FIGURE 9.12 Test suite using equivalence partitioning for the coming out roll of the game of craps.

as validly use a roll of 2 or 12, since they are in the same equivalence group as 3.

There is more than one way to slice a pizza. Some restaurants cut their pizza into wedges and others cut into squares. Equivalence partitioning also need not occur in just one way. Some testers might prefer to have five groups in their equivalence partitioning of the coming out roll by splitting the error group into two groups: one group for invalid rolls less than 2 and a second group for invalid rolls greater than 12.

The previous analysis considered equivalence partitioning for just the first (coming out) roll of a game of craps, but what if we wanted to write test cases around a complete game of craps? Figure 9.13 shows a test suite for an entire game.

Only one additional equivalence group is required. The prior group, called point, representing rolling something other than a 2, 3, 7, 11, or 12 on the coming out roll, has been replaced by two groups. The sevening out

Test Cases	Input Conditions	Expected Behavior
Crapping Out	coming out roll is 3	the user craps out
Natural Winner	coming out roll is 11	the user is a natural winner
Sevening Out	coming out roll is 4 followed by rolls of 3, 5, and 7	the user sevens out
Making the Point	coming out roll is 8 followed by rolls of 2 and 8	the user wins by making the point
Error	coming out roll is −4	the program reports an invalid roll

FIGURE 9.13 Test suite using equivalence partitioning for a complete game of craps.

group is the result of rolling a point in the coming out roll followed by subsequent rolls that produce a 7 before repeating the point. The making the point group establishes the point on the coming out roll, then eventually rolls the point again before rolling a 7.

9.7 BOUNDARY VALUE ANALYSIS

Boris Beizer, an author of books on software testing [1], says, "Bugs lurk in corners and congregate at boundaries." The point Beizer is trying to make is that software developers design programs to work for normal cases, which means that the less-than-normal cases are somewhat more likely to fail. This thinking leads to a black box technique for selecting test cases, known as *boundary value analysis* that is related to but also different from equivalence partitioning.

Using boundary value analysis on each input condition, we can select at least four different test cases:

1. One test case at a boundary

2. One test case just within the boundary

3. One test case just outside the boundary

4. One test case that is more typical (well within boundaries)

Notice that only the last test case is a typical—"middle value"—kind of test, whereas the others are intended to exercise input conditions near or at the boundary. Also, note that it is important to consider erroneous test cases to check, which explains why one of the cases is outside the boundary.

Often input conditions involving numeric values are specified as a range with a minimum value and a maximum. In such cases boundary value testing the numeric value results in seven test cases:

1. One test case at the minimum

2. One test case just greater than the minimum

3. One test case just less than the minimum

4. One test case at the maximum

5. One test case just less than the maximum

6. One test case just greater than the maximum

7. One test case somewhere between minimum and maximum (near the middle of the range)

For example, suppose we are developing the vehicle stability software for an automobile. This software is designed to prevent automobile rollovers so two of the important input conditions are the current speed of the car and the angle that the vehicle is leaning with respect to horizontal. Let's suppose that our software must operate whenever the automobile is traveling in the range from 0 mph to 150 mph. Using boundary value analysis our test suite should include at least seven test cases, such as those from Figure 9.14.

When there are multiple input conditions, the number of test cases required for complete boundary testing grows rapidly. Returning to our vehicle stability software example, assume that the permissible range of lean is −45° to 45°. Exhaustive testing of all boundaries entails all combinations of conditions. Figure 9.15 depicts this as a two-dimensional grid consisting of the boundary value analysis for vehicle speed on one grid axis and the boundary value analysis for vehicle angle of lean on the other axis. Each dot on the grid represents a separate test case. The upper-left dot is for a speed of 151 and a angle of −46°; the second dot in the top row is for a speed of 151 and angle of −45°; and so forth. The total number of test cases for such testing is 7 times 7 or 49.

As another example of boundary value testing consider the calculation of income tax. Figure 9.16 shows a table of United States income tax

Test Cases	Input Conditions	Expected Behavior
At Minimum	vehicle speed = 0 mph	car doesn't move
Minimum +1	vehicle speed = 1 mph	car doesn't roll over
Minimum −1	vehicle speed = −1 mph	car in reverse with no rollover avoidance
At Maximum	vehicle speed = 150 mph	car doesn't roll over
Maximum −1	vehicle speed = 149 mph	car doesn't roll over
Maximum +1	vehicle speed = 151 mph	car issues a *driving too fast* warning
Typical	vehicle speed = 70 mph	car doesn't roll over

FIGURE 9.14   Test suite for vehicle stability software considering only vehicle speed.

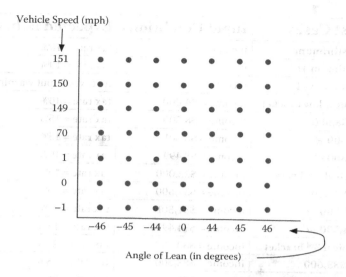

FIGURE 9.15   Boundary value analysis for both speed and angle of lean.

brackets for single individuals from 2000 through 2011. We can see more boundaries in this kind of table, one for each tax bracket separation. There are a total of 24 test cases as shown in Figure 9.17. In this case there is no particular maximum for the top bracket, and its "typical" is taken to mean just some representative amount well within the range.

The input conditions examined thus far have involved numeric input where boundaries are often expressed as minimum and maximum allowable values. For other types of input the boundaries might be less obvious. For example, suppose the input conditions involve textual input. In this case we could look for boundaries as the extremities. The minimum and maximum length for the text can be thought of as boundaries. In some

Taxable Income	Tax Rate
$0 – $8,500	10%
$8,500 – $34,500	15%
$34,500 – $83,600	25%
$83,600 – $174,400	28%
$174,400 – $379,150	33%
$379,150 and above	35%

FIGURE 9.16   US federal tax brackets for single individuals (2000–2011).

Test Cases	Input Conditions	Expected Behavior
At Minimum	income = $0	tax rate = 10%
Minimum +1	income = $1	tax rate = 10%
Minimum −1	income = −$1	no taxes (input warning)
Typical low bracket	income = $4,000	tax rate = 10%
At $8,500	income = $8,500	tax rate = 15%
$8,500 +1	income = $8,501	tax rate = 15%
$8,500 −1	income = $8,499	tax rate = 10%
Typical 2nd bracket	income = $22,000	tax rate = 15%
At $34,500	income = $34,500	tax rate = 25%
$34,500 +1	income = $34,501	tax rate = 25%
$34,500 −1	income = $34,499	tax rate = 15%
Typical 3rd bracket	income = $60,000	tax rate = 25%
At $83,600	income = $83,600	tax rate = 28%
$83,600 +1	income = $83,601	tax rate = 28%
$83,600 −1	income = $83,599	tax rate = 25%
Typical 4th bracket	income = $130,000	tax rate = 28%
At $174,400	income = $174,400	tax rate = 33%
$174,400 +1	income = $174,401	tax rate = 33%
$174,400 −1	income = $174,399	tax rate = 28%
Typical 5th bracket	income = $275,000	tax rate = 33%
At $379,150	income = $379,150	tax rate = 35%
$379,150 +1	income = $379,151	tax rate = 35%
$379,150 −1	income = $379,149	tax rate = 33%
Typical top bracket	income = $1,000,000	tax rate = 35%

FIGURE 9.17   Test suite for income tax brackets.

cases the context of the text (i.e., which symbols are allowable) might be used to determine extreme conditions.

9.8 WHEN WILL YOU EVER USE THIS STUFF?

Admittedly, few of us will ever need to prove that a computer program is correct, but as this chapter points out neither do most computer scientists. (Although, it is comforting to know that such proofs are possible and can be used in cases of life-endangering software.)

More important, the fact that software testing does not ensure correctness explains why virtually all computer programs contain flaws.

Although modern software is pretty reliable due to extensive and effectively designed testing, you should never be lulled into total confidence when it comes to computer programs.

The software testing process is also broadly applicable to testing almost any product. A toy company could test toy assembly directions by treating these directions as an algorithm and using the concepts of statement coverage or path coverage.

Any supervisor can benefit by applying boundary value analysis to employee evaluation. The basic idea is that while you should test using a typical situation, focusing even more on the extreme (boundary) conditions tends to ensure thorough testing. Every camera manufacturer knows that it is more difficult to build a camera lens that focuses perfectly around the edges than one with center sharpness.

You could even apply your knowledge of testing to the purchase of a new car. You have learned that there can be no correctness without specifications. So the car purchase process needs to begin with a list of requirements; we could make a list of requirements such as minimum acceptable gas mileage, acceptable colors, 0 to 60 time, cost, and so forth. Next we create a test suite of test cases. One test case might be to visit the EPA website to determine mileage; another case could use Kelly Blue Book information to determine resale value; another test case might involve handling evaluation from a test drive. The point is that we can apply computer science testing concepts to almost any decision. The more important the decision, the more extensive should be the testing strategy.

REFERENCE

1. Beizer, B. *Software Testing Techniques*. Boston: International Thomson Computer Press, 1990.

TERMINOLOGY

acceptance testing	customer
beta testing	equivalence partitioning
black box testing	flaw (software)
boundary value analysis	functional testing
bug	hardware failure

improper output interpretation	specifications (for software)
incorrect data entry	statement coverage
layering	structure testing
nonfunctional requirement	system misunderstanding
path testing	test case
physical (computer) damage	test suite
proof of correctness	usability testing
regression testing	V&V
security breach	validation
software assurance	verification
software fault	white box testing
software testing	

EXERCISES

1. Identify for each of the following whether the procedure described is more like a verification process or a validation process.

   a. You work for an accounting firm and audit all of the checking accounts for a particular bank.

   b. You work for an architectural firm and specialize in showing initial plans to customers to get their input.

   c. You work for the government and test food products to ensure that their actual ingredients are the same as printed on the product label.

   d. You are a writer for a television sitcom and need to convince the show's producers that your jokes are consistent with the intended personalities of the characters.

2. If you are the customer for a software application, which of the following kinds of testing would you be least likely to perform?

   a. Acceptance testing

   b. Beta testing

   c. Black box testing

   d. White box testing

3. A pharmaceutical company writes contracts with its chemical suppliers that ensure each container has the proper amount of chemical within some allowable range. For example, a one liter container is permitted to contain between 997 cm^3 and 1,003 cm^3. Design a test suite for testing the company's quality control using a liter container of a different amount of chemical for each different test case.

4. Suppose a vending machine company designs a new machine that will dispense anything that is stored in a plastic bag, similar to a bag of potato chips. The ideal size for a bag dispensed by this machine is 11 cm by 18 cm, but the machine is designed to accept bags that are 1 cm longer or shorter in either dimension. Design a test suite for this machine using the concept of boundary value analysis.

5. Imagine that you are designing a knockoff of the popular Angry Birds game. A simple version of your game will have three different kinds of birds (round, triangular, and square) along with two kinds of materials for building structures around the pigs (wooden or steel). Each kind of bird behaves differently when striking material. Each material produces different results when struck by a bird, except for round birds, which behave the same for both wood and steel. Describe the different equivalence classes of this system in terms of bird and material type. You should select each group in such a way to ensure that it has a behavior that is different from all others.

6. Design a test suite for the following algorithm for an online account to manage a stock portfolio. Number the activities in the diagram and build a table to demonstrate that your test suite achieves statement coverage.

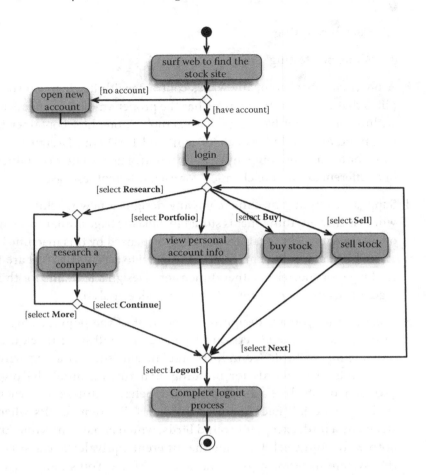

CHAPTER 10

Limits of Computation

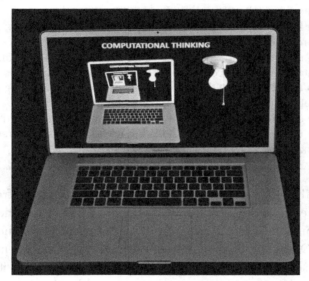

The 9000 series is the most reliable computer ever made. We are all, by any practical definition of the words, foolproof and incapable of error.

—**HAL**

The computer in the movie 2001: A Space Odyssey[1]

OBJECTIVES
- To recognize that computers are not, and will never be, foolproof or limitless in their problem-solving potential
- To recognize that computing capacity is a two-part issue, consisting of storage capacity and processing speed, with processing speed being the more important

- To understand that Moore's law has been achieved to date largely because of miniaturization, but multicore processors are contributing recently; and to be aware that these improvements cannot continue forever
- To realize that benchmarks can be used to measure a computer's ability to solve problems, but that each benchmark provides only one highly specific data point that may or may not be generalized
- To recognize the common linear algorithms as demonstrating a proportional growth that can be graphed as a line
- To know that a binary search proceeds by repeated selecting from the middle of the data, and that this algorithm works only for sorted data but significantly outperforms a linear search
- To know that algorithms with polynomial performance are generally considered to be tractable for computer processing
- To recognize that exponential algorithms, except for small amounts of data, are generally considered to be impractical for computer execution
- To understand that there are problems, such as the halting problem, that can never be solved by a computer program
- To explain the Turing test and how it relates to computer intelligence
- To understand how CAPTCHAs are useful and that they represent a kind of reversal of the Turing test

Arguably the most famous computer quotation of all time is from Stanley Kubrick's epic motion picture *2001: A Space Odyssey* [1]. The computer in this film, named HAL, proclaims itself to be "foolproof and incapable of error." Of course, there have been countless other stories, movies, and television shows that have portrayed computers as devices with limitless capabilities. We humans seem truly fascinated by the thought that our inventions might one day be infallible in ways that we are not.

But are computers with unlimited abilities a possibility or merely science fiction? In this chapter we explore what is known today about the limits of computation. We shall explore the concept from the perspective of physical limitations, algorithmic limitations, and logical reasoning, and also briefly look at the metaphysical.

In the very first chapter it was mentioned that one of the factors that makes computers so remarkable is the rate at which computer-related technology has advanced. Up to this point in our history, the processing capabilities of computer hardware has largely followed *Moore's law*—doubling every 18 months. This means that today's computers are more than 100 times faster than computers 10 years of age and over 10,000 times faster than those from 20 years ago.

Time Passed	Increased Capability
0 yrs	x1
1.5 yrs	x2
3 yrs	x4
4.5 yrs	x8
6 yrs	x16
7.5 yrs	x32
9 yrs	x64
10.5 yrs	x128
12 yrs	x256
13.5 yrs	x512

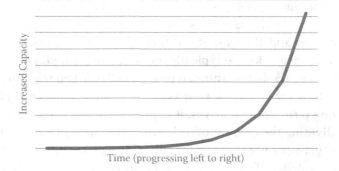

FIGURE 10.1   The exponential growth of Moore's law.

Such a rapid change as Moore's law is often called *exponential growth*. Figure 10.1 contains a table and graph that illustrates this. Exponential growth occurs whenever a function increases by compounded powers. In the case of Moore's law we can think of the base as two for a doubling and the power as the number of 18-month units in the passing time. The characteristic of exponential growth that is worthy of note is that it increases at an ever-growing amount, as shown by the graph in Figure 10.1. Take note of the fact that the plotted function curves upward by more and more as time passes.

Certain microorganisms also exhibit exponential growth for a period of time. This happens when one cell splits to form two, each of the two split to form four, the four split to make eight, and so forth. In the natural world exponential growth is almost always limited by some constraint. If we are growing cells in a petri dish, then the constraint might be the size of the dish or perhaps the limited amount of chemicals needed to sustain the reaction. Again, this leads to the question: Is Moore's law sustainable for computing devices?

10.1 HOW IS CAPACITY MEASURED IN COMPUTERS?

We measure the speed of our automobiles in miles (or kilometers) per hour. We measure an amount of soda in ounces or milliliters. Perhaps before we explore the limitations of computing capacity, it would be important to understand just how such things are measured.

The question of capacity is thought of as a two-part question: (1) how to measure processing speed and (2) how to measure storage size. Computer scientists know this as an issue of *time and space.*

Space (storage capacity) is the easier thing to measure. Every computer manufacturer clearly advertises the amount of space in their computers' memories and the capacity of their disks. As discussed in Chapter 2, computer memories are typically measured in gigabytes and disk sizes in gigabytes or terabytes. This can be instructive if you want to know, for example, how many photos your computer can retain. Since photos are stored long term on disk drives, it is possible to estimate the photos per disk by dividing the amount of disk storage by the typical size of a photo. Suppose that your photos are generally 2 megabytes in size and that you have 100 gigabytes of unused space on your computer's disk drive. In this case you can estimate that approximately 100 gigabytes/2 megabytes = 51,200 more photos can be stored, assuming that the disk is used for nothing else. (Recall that one gigabyte equals 1,024 megabytes.)

Memory space almost always greatly exceeds the size of the item being processed. For example, an employee record or a graphical image might occupy a few kilobytes, whereas your computer's memory is a few gigabytes (hundreds of thousands of times larger). Even a video of say 20 minutes is probably compressed into a space of 100 megabytes to 1 gigabyte, which is easily smaller than a common laptop computer memory of 2 to 8 gigabytes.

The truth is that you do not purchase computer memory so much based on the size of the things being stored, as much as to improve the speed of processing those things. It turns out that software is typically more efficient when it has greater available memory. Think about rearranging the furniture in your bedroom. If you have empty space in which to temporarily relocate the furniture pieces, it is easier and faster to move them. Computer data often behaves in the same way—more space (computer memory) generally allows the software to manipulate your data faster. Of course, there are diminishing returns. A 3,000 square foot shed is not really needed in order to rearrange one bed and a small nightstand.

Relative to other parts of a computer, the storage units—both memory and disk—tend to be inexpensive. Especially memory prices per unit have followed, and often outpaced, Moore's law. In part for this reason, most of today's consideration of computing capacity falls into the time (processing speed) category.

Estimating time performance is a difficult task. Suppose you are responsible for selecting new computers for your company. Given several options of similar cost, you might want to choose the computer that is fastest. The challenge is how to measure "fast."

Computer manufacturers publish the speed of their processors using a *clock rate*. Inside each processor is a clock that cycles many times per second. A modern processor might have a clock rate of 1 to 3 *gigahertz* (GHz). One gigahertz equals one billion cycles per second, and higher numbers translate into more frequent cycles. If you compare two processors of the same type, then the one with the higher clock rate can be expected to be faster.

Unfortunately, there are many different kinds of processors, so one type of processor with a higher clock rate may or may not be faster than a different type of processor with a slower clock rate. The reason for these discrepancies is twofold:

1. Different types of processors vary in the number of clock cycles per instruction.

2. Different types of processors have different kinds of instructions.

So the clock rate is an imperfect measure of how much work the computer can perform, both in terms of number of instructions it can execute and the power of each instruction.

In an attempt to improve the measure of speed, companies sometime quote processor speed in terms of millions of instructions per second (*MIPS*) or millions of floating point operations per second (MFLOPS).

MIPS and MFLOPS are somewhat adjusted for differences in instruction types, so this kind of measure is more instructive than simply a clock rate. In fact the world's fastest computers, the so-called *supercomputers*, are most often compared using MFLOPS.

To further complicate the issue, it is common to have more than one processor in a computer. The term *core* is often used as a way to differentiate the processor count. If you own a *duo-core* computer, then your computer has effectively two primary processors and a *quad-core* computer has four. Obviously, multiple processors are capable of more calculations than the same type of computer with fewer processors. In addition, our computers contain graphics processors, computer memories, and disk drives—all of which effect the overall computer speed. A faster processor, for example, can be restricted by a slower memory.

10.2 AN ESTIMATE OF THE PHYSICAL LIMITATIONS

Modern computers are mostly constructed from electronic components. These devices have followed Moore's law primarily due to *miniaturization*. As electrical circuits have become smaller, they also have become faster.

The particular kind of technology that has fueled this miniaturization is *semiconductors*. Semiconductor components store and manipulate electrons, and those electrons flow from one component to another on narrow paths. The devices and paths are most often etched by laser beams that are as small as a few thousandth of the thickness of a human hair. Creating smaller and smaller devices (and narrower laser beams) is becoming increasingly difficult. Further, there is an obvious end to potential miniaturization: the size of an electron. Just like you could not drop a marble into a pipe unless the diameter of the pipe is greater than the marble, so too the electron paths must be greater than the width of an electron. Fortunately, for now, the diameter of an electron is millions of times smaller than current technology limitations. However, Moore's law will reach this limit more quickly than you might imagine.

There are at least three other technologies that have been considered as future ways to fabricate computers:

1. *Optical computers*

2. *Biological computers*

3. *Quantum computers*

	Lloyd's Ultimate Computer	Today's Laptop Computer
Processor (operations/sec.)	10^{32}	10^{9}
Memory Size	10^{16} bits	10^{10} bits

FIGURE 10.2   Lloyd's ultimate computer limits compared to a laptop.

The basic idea of an optical computer is to store bits as light rather than electricity. Bits can be represented as photons. One advantage that has been demonstrated by early research is that an optical memory can be three dimensional, whereas electronic memories are essentially two dimensional.

Some scientists believe that DNA can be used as a computational device. Perhaps the resulting biological computers will be built molecule by molecule, but actual devices of this sort are not on the immediate horizon.

Quantum computers, on the other hand, have already been constructed in a few early research labs. Based on quantum theory that explains an electron's location as probabilistic, quantum computers are expected to store data in more complex forms than simple bits. This could result in a more compact form of data.

All three of these new technologies are best characterized as being in their infancy. Furthermore, all are limited, admittedly in different ways, by the size of elementary particles.

An interesting analysis of the limits of computing devices was published in the year 2000 by a physicist named Seth Lloyd [2]. Lloyd describes something he calls the "ultimate computer" based upon the laws of physics (Figure 10.2). His "computer" would consist essentially of a solid cube of dense matter. Without regard for the way that such a computer might function, but only the potential for the behavior of dense three-dimensional matter, Lloyd concludes theoretical limits of 10^{32} operations per second and data in the range of 1,016 bits.

Although it is impossible to know whether computer technology will ever permit anything close to the limits described by Lloyd, it is clear that Lloyd's limits allow for at least another 100 years of Moore's law advances. Realistically, experts suggest that the limit for continued advances at the rate of Moore's law are closer to a decade or two.

10.3 BENCHMARKS

So far this chapter has primarily concentrated on the capabilities of computer hardware, memory size and processor speed in particular. But the true measure of a computer is its utility in problem solving and this requires an analysis of both hardware and software. There are two well-known ways to analyze such combined capabilities, one empirical and one analytic.

Thinking empirically is a kind of scientific investigation that is based upon experimentation. Thinking analytically relies upon logical reasoning from known facts to draw inferences without an actual experiment. In this section we examine an empirical approach, leaving the analytic for later in the chapter.

The empirical approach is similar to using a stopwatch to time a track runner, except the thing you are timing is a computer application running on some particular computer. Such an application timing is known as a *benchmark*.

For example, you could benchmark a computer and program for completing your income tax by executing the program. There are several benefits of benchmarks. They permit testing for characteristics other than speed. If we used last year's data for our benchmark of the income tax program, we could check the program for correctness. Sometimes benchmarks are used to determine such things as

- Response time—How quickly does the application respond to user input?

- User friendliness—How easy is the application to use?

- Robustness—How immune to failure is the software?

Another benefit of benchmarks is that they analyze both hardware and software, presumably in a combined way that resembles the intended purpose. Furthermore, a benchmark can sometimes be meaningful in comparing two computers by using the same software and data to benchmark each computer. Similarly, two different computer programs for the same application can be compared using two separate benchmarks.

The major shortcoming of benchmarks is that they are difficult to generalize. A benchmark measures only a single set of circumstances. Just because the application behaved well for one year's income tax does not mean it will do so for a different year, different deductions, or different tables. In addition, the timing of a benchmark can change the outcome substantially. Were you playing music on your computer at the same time as the benchmark execution? Unbeknown to you, was your computer downloading e-mail or scanning for viruses during the benchmark?

Good science requires experiments that are repeatable with carefully controlled variables. Yet, both repeatability and control of benchmark impacting variables are nearly impossible with computer benchmarks. Modern software generally allows for a vast number of potential inputs, and modern operating systems execute multiple applications simultaneously with little or no user control. About all that can be concluded with absolute certainty from many benchmarks is that this was the behavior given the computer's state at that time and given that particular set of user inputs. Sometimes even processing just one additional deduction on your income tax can cause the application to behave in significantly different ways.

Despite the drawback of benchmarks, this technique can still be useful. Techniques for improving the usefulness of benchmarks include performing multiple benchmarks and using inputs as close as possible to those of the intended use.

10.4 COUNTING THE PERFORMANCE

One of the most challenging issues for benchmarking is to recognize differences in performance that are dependent upon the volume of data. Computers tend to be particularly good at processing large numbers of things, such as interacting with hundreds of simultaneous online website customers, or handling thousands of credit card transactions, or determining the result of wind blowing across millions of hairs on a dog's nose in a motion picture animation. Unfortunately, one benchmark only considers a single data volume.

Another performance estimation technique used frequently by computer scientists relies on the analytic approach of counting tasks in an algorithm. We could count the number of page hits per customers making an online purchase; we could count the number of additions and subtractions during each credit card transaction; we could count the number of image changes while animating a video game avatar. Ideally, computer scientists count the number of computer instructions executed or the number of times data is retrieved from or stored in memory.

We can apply these same counting techniques to estimate the time required to perform everyday tasks. For example, suppose a carpenter wishes to estimate the time required to install trim boards in a house. The carpenter observes that the men attaching the boards perform this job in about the same length of time as the man who measures then saws the boards to the correct length. Therefore, a reasonable approximation of work time can be made counting how many boards need to be

```
holdTicket ← first ticket from the stack
remove holdTicket from stack
while  stack not empty AND holdTicket not winner  do
    holdTicket ← next ticket
    remove holdTicket from stack
endwhile
if  holdTicket is winner  then
    The winning ticket is holdTicket
else
    The stack contains no winning ticket
endif
```

FIGURE 10.3 Algorithm A for searching for the winning ticket.

sawn. The number of boards may not translate exactly into time, because one carpenter may be consistently faster than another. However, we can draw useful conclusions about the speed of any given carpenter and can usefully infer the impact of differing board counts on the overall project timing.

Similar counting is also useful for comparing the speed of different algorithms. For example, suppose that you are given a stack of lottery tickets, each containing a different lottery number. You are asked to examine the tickets to find out if any of the tickets contains the winning number. One algorithm to search for the winning ticket is described in Figure 10.3; we will call this Algorithm A.

The portion of this algorithm that gets repeated consists of removing from the stack the previously assigned holdTicket and assigning the next ticket in the stack to holdTicket. Therefore, we could estimate the speed of this algorithm by counting how many tickets need to be examined. There are two possible cases to consider: (1) if the stack contains no winning ticket and (2) if there is a winning ticket in the stack. In the case that the tickets are all losers, then the number of tickets to examine is identical to the number in the stack. (For a stack of 10 tickets, we must examine all 10; for a stack of 20, we examine 20; and so forth.)

The performance of Algorithm A when the stack contains a winning ticket is more difficult to count, because the number of tickets that must be examined depends upon where the winning ticket is located within the stack. Without any additional information we might conclude that the winning ticket is equally likely to be anywhere, so probability would suggest that it is in the middle. If the ticket is precisely in the middle, then the speed of the algorithm can be counted as half the number of tickets in the stack.

Number of Tickets in Stack	Stack contains winning ticket	No winning ticket in stack
0	0	0
10	10	5
20	20	10
30	30	15
N	N	N/2

FIGURE 10.4   Count of tickets examined for Algorithm A.

Figure 10.4 contains a table that analyzes these counts in more detail. This table shows the count of tickets that require examination for ticket stacks of size 0, 10, 20, and 30 tickets. The bottom table row is a useful generalization that shows the count based upon a variable (N) number of tickets. In other words, given a stack of N tickets, no matter of the value of N, the number of examinations will be either N or N/2.

Regardless of whether the stack contains the winning ticket, we conclude that the performance of this kind of search algorithm is *directly proportional* to the number of tickets in the stack. Computer scientists have a name for algorithms that perform according to a direct proportion; they are called *linear algorithms*.

To see where the name "linear" comes from, consider the graph of the performance of Algorithm A, as shown in Figure 10.5. The red solid line in this figure diagrams the algorithm when no ticket is found, and the dashed green line shows the performance when the ticket is found in the

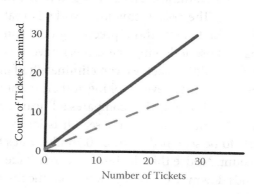

FIGURE 10.5   Performance of Algorithm A.

```
while  the stack has more than one ticket  do
    Examine the middle ticket;
    if middle ticket number less than winning number  then
        Remove from stack everything through middle ticket.
    else
        Remove from stack everything following middle ticket.
    endif
endwhile
if  a ticket remains in the stack and is the winner  then
    The winning ticket is the one left.
else
    The stack contains no winning ticket.
endif
```

FIGURE 10.6   Binary searching for the winning lottery ticket.

middle of the stack. Notice that both of the graphed counts form a line, and so they are considered linear algorithms. For this reason this algorithm is known as a *linear search*.

Linear algorithms are probably the most common in everyday life, because we often assume that if processing one piece of data requires a certain amount of time, then processing N pieces of data will require N times more. However, there are a few rare algorithms that actually perform better.

Suppose we know that our stack of lottery tickets is sorted in order from lowest number to highest. In this case we can employ a more efficient algorithm, namely, the *binary search* algorithm previously examined in Chapter 4.

Figure 10.6 describes this algorithm. The key discovery in a binary search is to repeatedly examine the middle ticket, rather than a ticket from one end of the stack. The reason this idea works is that from the middle roughly half of the tickets (those preceding the center) have smaller numbers and half (those following the center) have larger numbers. By examining only the middle ticket, we can eliminate half of the stack from further consideration without even looking at their numbers!

Figure 10.7a contains a table that compares a linear search to a binary search. The general formula for binary search is $\log_2 N$, where N is the number of tickets to be searched. Figure 10.7b illustrates the significance of this kind of count. Notice that the binary search (blue dashed) performance curve bends downward moving away from the linear search line as the number of tickets grows.

Number of Tickets in Stack	Linear search	Binary search
	0	0
4	4	2
8	8	3
32	32	5
N	N	$\log_2 N$

(a)

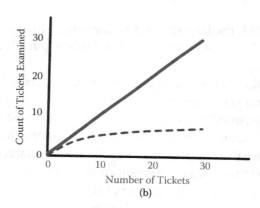

(b)

FIGURE 10.7 (a) Count of tickets examined for Algorithm A. (b) Performance of binary search compared to linear.

By counting in this way we can conclude that given the same hardware a computer program using a binary search executes faster than a similar program using linear search, especially when searching through large amounts of data.

Although binary search has a better performance than linear search, there are other algorithms that execute slower than linear search. To examine this possibility, we examine a different task, that of sorting a random stack of lottery tickets so that their numbers are ordered from smallest to largest.

One algorithm for sorting is to repeatedly find and remove the largest ticket from the unsorted stack by moving each ticket with the largest number from the unsorted stack to a new, sorted, stack. The first number removed would be the largest of all tickets. Once this largest

```
holdTicket ← first ticket from the stack
remove nextTicket from stack
while  the stack is not empty  do
   nextTicket ← next ticket from the stack
   remove nextTicket from stack
   if  nextTicket > holdTicket  then
      holdTicket ← nextTicket
   endif
endwhile
// holdTicket has greatest number from the stack
```

FIGURE 10.8   Algorithm to find the largest number in a stack of lottery tickets.

ticket is removed, the largest ticket in the remaining stack must be the second largest from the original, and so forth. Figure 10.8 summarizes this algorithm.

Let's count the number of times that a ticket needs to be moved from one stack to another. Since every ticket is moved exactly once from the unsorted to the sorted stack, the number of moves must be the same as the number of tickets. It might look like this is another linear algorithm, unless you notice that the process of searching for the largest ticket is itself a repeated effort that takes a significant amount of time. Figure 10.9 details this algorithm for finding the largest ticket.

Instead of counting how many tickets need to be moved, a more accurate count of the Figure 10.9 algorithm's speed comes from counting how many times we need to examine each ticket. A combined analysis of the algorithms from Figures 10.8 and 10.9 reveals that given N tickets, searching for the largest ticket requires N tickets be examined. Searching for the second largest occurs in a stack that already has a ticket removed, so this second largest is found by N − 1 examinations. Similarly, the third largest is found in N − 2 examinations and the fourth in N − 3. Therefore, we can count the total number of tickets examined to be N + (N−1) + (N−2) + (N−3) + ⋯ + 1. See Figure 10.10.

```
Begin with a stack that is sorted & empty.
while  the stack contains one or more tickets  do
   Find largest numbered ticket in unsorted stack.
   Move found ticket to top of sorted stack.
endwhile
```

FIGURE 10.9   Algorithm for sorting a stack of lottery tickets.

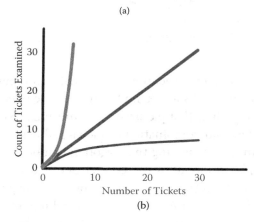

$$\text{Total tickets examined} = \text{tickets examined to find largest} + \text{tickets examined to find 2nd largest} + \text{tickets examined to find 3rd largest} + \cdots + \text{tickets examined to find last largest}$$

$$\text{Total tickets examined} = N + (N-1) + (N-2) + \cdots + 1$$

$$\text{Total tickets examined} = N^2 - \frac{N}{2}$$

(a)

(b)

FIGURE 10.10   (a) Compares the performance of all three algorithms. The top (green) curve is for our sorting algorithm, while the binary search is the bottom (blue) curve and the linear search is shown in red between. (b) Performance of sorting, linear search and binary search.

10.5 IMPRACTICAL ALGORITHMS

The sorting algorithm from the previous section points out that there are algorithms that perform worse than linear algorithms. But how much worse can they be? And if they are worse does their performance impose any kind of limit on computation? These are the questions explored in this section.

All of the algorithms examined in the previous section can be described as having a *polynomial performance*, because each formula for counting in terms of the number of data items (N) is a polynomial. Broadly speaking, polynomial algorithms are considered to be appropriate computations for computers. In fact most computer programs have performance that can properly belong to the polynomial performance category.

FIGURE 10.11 Combination padlock.

Some algorithms have execution times that grow much faster than polynomial algorithms. As an example consider the operation of a combination padlock such as that pictured in Figure 10.11. Now suppose that you have forgotten your combination to such a padlock.

One algorithm for recovering the forgotten combination is to attempt all possible combinations until the correct one is discovered. What we know about padlocks of this type is that a combination generally consists of rotating the dial clockwise to one integer marking, then counterclockwise past and to a second integer marking, followed by a clockwise rotation to a third integer marking. So the process can be represented as a sequence of three integers, potentially including duplicates. By examining the particular padlock from Figure 10.11 we can also observe that there are 40 possible integer markings (0 through 39) for each rotation. This means that the total number of potential combinations for this padlock is calculated as follows:

$$40 \times 40 \times 40 = 40^3 = 64{,}000 \text{ combinations}$$

Another way to think about this is that an algorithm to discover the combination we would need to perform $40^3 = 64{,}000$ different tests in the worst case. If it takes 5 seconds to attempt a combination, it would require $64{,}000 \times 5$ seconds, or roughly 90 hours, in the worst case. Presumably, the padlock manufacturer has decided that 90 hours is sufficiently secure to discourage criminals.

Number of Markings (m)	Number of Combinations (m³)	Time (at 5 sec. per combination)
40	64,000	88.89 hours
41	68,921	95.72 hours
42	74,088	102.90 hours
43	79,507	110.42 hours

FIGURE 10.12   Combinations and time required to discover a combination for different numbers of dial markings.

However, if we want to build a combination padlock that is more secure there are two obvious choices:

1. Include more numbered markings around the dial

2. Require more than three rotations

It turns out that including more numbers around the dial, in addition to making the padlock more difficult to operate, exhibits a polynomial algorithm performance. In other words, as the number of markings increases, the number of combinations increases according to a polynomial formula. We can describe the worst-case performance as m^3 combinations, where m is the number of markings (Figure 10.12).

The second way to build a more secure padlock is to increase the number of rotations. For example, instead of three rotations to three number markings (clockwise–counterclockwise–clockwise) we could choose to require four rotations to four number markings (clockwise–counterclockwise–clockwise–counterclockwise). If four rotations are insufficient, then require five or more. This leads to a performance that varies in terms of the number of rotations. If we symbolize the number of rotations with the letter r, then there are 40^r combinations to test in the worst case. Because the variable (r) is an exponent, this kind of algorithm is commonly called exponential. Not surprisingly, an algorithm whose performance is described by an exponential formula is known as an *exponential algorithm*. Figure 10.13 is a table that demonstrates this kind of change.

Number of Rotations (r)	Number of Combinations (40^r)	Time (at 5 sec. per combination)
3	64,000	88.89 hours
4	2,560,000	296.29 days
5	102,400,000	16.22 years

FIGURE 10.13 Combinations and time required to discover a combination for different numbers of rotations.

The tables from Figure 10.12 and 10.13 illustrate the considerable difference between polynomial growth and exponential growth. Increasing the number of markings (m) by one means that the algorithm for discovering a combination (m^3) takes a few hours more to complete. By comparison, increasing the number of rotations (r) by one means that discovering a combination (40^r) goes from hours to days or from days to years. We could graph these differences as shown in Figure 10.14. Changing the number of markings only (the polynomial algorithm) is diagrammed with a solid red line and changing the number of rotations only (the exponential algorithm) appears as a green dashed line. The difference is so extreme that the exponential graph is nearly vertical, whereas the polynomial algorithm is nearly horizontal.

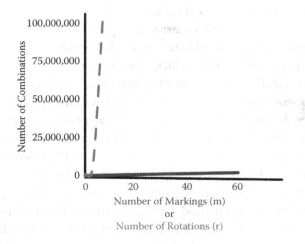

FIGURE 10.14 Graphing the polynomial and exponential algorithms for different kinds of padlocks.

Amount of Data (D)	D³	3ᴰ
1	1 sec.	1 sec.
5	2.08 min.	4.05 min.
10	16.67 min.	16.40 hr.
15	56.25 min.	166 days
20	2.22 hr.	110 years
25	4.34 hr.	26,849 years
30	7.50 hr.	6,524,296 years
35	11.91 hr.	1,585,403,995 years
40	17.78 hr.	385,253,170,680 years*

*This number is roughly 26 times the age of the universe by current estimates.

FIGURE 10.15   Comparison of D^3 and 3^D growth.

The extreme time growth that is demonstrated by exponential algorithms can be illustrated more simply by considering two simpler algorithm performance formulas. Let's suppose for some amount of data, called it D, that one algorithm requires D^3 seconds (polynomial growth) to execute, while a second algorithm requires 3^D seconds (exponential growth) to perform the same task. The table in Figure 10.15 compares the execution time of each algorithm for different amounts of data. Not surprisingly, both algorithms need a longer time period to execute more data, but the difference between the algorithms is extreme.

From Figure 10.15 we see that this exponential algorithm takes longer to execute than the presumed age of the universe for data amounts greater than 40. This algorithm will simply never finish running for any large amount of data. Even if our computers continue to become faster at the rate of Moore's law, the gains will never support reasonable execution times for exponential algorithms and large amounts of data.

The typical kinds of algorithms that are exponential, and therefore impractical, are brute force solutions in which a small increase in data causes the execution time to multiply. Algorithms that attempt all possible combinations or all possible permutations tend to be exponential. Examples of such algorithms would be a computer program to analyze all possible moves in a chess game or all possible genetic sequences.

It is the impracticality of exponential algorithms that allows us to securely use passwords. A simple algorithm to crack any password that could be written by a first-year computer science student is the brute force approach of attempting all possible character strings. If we consider a keyboard of 96 keys and assume that we know the password is of length P, then the number of potential passwords is 96^P. This means that our brute force password cracking program is exponential in terms of the password length. Therefore, if we choose a password length that is sufficiently large (around 11 or 12 symbols is enough), then a brute force algorithm is essentially useless.

One last caution about exponential algorithms: Just because one algorithm for a problem is exponential, it does not mean that there is not a better algorithm. For example, it is possible to write a program that cracks a password in seconds, assuming the password is an English word. Rather than the brute force algorithm, this program would test all words from an English dictionary. The number of possible English words is around 300,000 to 500,000, which can be checked in seconds or less. This is why you are told not to use a password that is commonly available in any dictionary.

10.6 IMPOSSIBLE ALGORITHMS

In Section 10.5 we discussed algorithms that are too impractical for use in computer programs. In this section the exploration of algorithmic limits leads from the impractical to the impossible. Computer scientists use a classic example known as the *halting problem* to illustrate impossible algorithms.

To understand the halting problem it is important to observe that some algorithms will execute forever, assuming there is no manual interruption. For example, consider the following while instruction:

```
while true do
    // do something
Endwhile
```

Since the condition of this loop (true) is always true, the loop continues to repeat its body forever. Computer scientists use the name *infinite loop* to refer to such a segment of code that never stops repeating. If this code is executed, then it will never stop executing the loop body, or in other words it will not halt.

The Halting Problem begins by making a key assumption. We assume that it is possible to create an algorithm (a computer program) to check

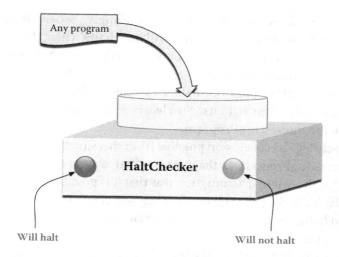

Will halt **Will not halt**

FIGURE 10.16   HaltChecker.

whether any given computer program halts when it is executed. Let's call the assumed algorithm HaltChecker.

Think of HaltChecker as a computer program that accepts as input another program. (Perhaps this other program is read from a file.) The purpose of HaltChecker is to determine whether the other program will halt.

Figure 10.16 depicts HaltChecker as black box machine. Another program, labeled "any program" in the figure, is input to the box after which the algorithm in the box lights one of two lights ("will halt" or "will not halt").

Now consider the following program; call it InterestingProgram:

```
if  HaltChecker says InterestingProgram will halt  then
    while  true  do
        // do something
    endwhile
else
    // halt
endif
```

InterestingProgram begins by invoking HaltChecker. The input for this HaltChecker invocation is the InterestingProgram itself. Of course, HaltChecker will return one of two results: either it determines that InterestingProgram will halt or that InterestingProgram will not halt.

Now consider the two cases. If HaltChecker returns "will halt," then this causes the InterestingProgram code to execute its while loop, which does not halt. Correspondingly, when HaltChecker returns that InterestingProgram "will not halt," then the aforementioned code executes the else option and halts. In other words, when HaltChecker determines that InterestingProgram will halt, then InterestingProgram does not halt; and when HaltChecker determines it will not halt, then it halts. So clearly HaltChecker does not work properly for InterestingProgram.

The only logical conclusion possible from this situation of HaltChecker and InterestingProgram is that our original assumption was incorrect. Recall that the original assumption was that it is possible to write an algorithm (HaltChecker) that given any program can determine whether the program halts. Therefore, we conclude that it is not possible to write such an algorithm.

To be a little more careful, the preceding analysis demonstrates that it is not possible to create a HaltChecker algorithm that works all of the time. However, it might still be possible for HaltChecker to work some, or even most, of the time. Nonetheless the preceding argument serves to unequivocally prove that there are certain algorithms, such as HaltChecker, that are impossible. So no computer program will ever be able to solve the halting problem.

It is interesting to note that a mathematician discovered the halting problem even before the modern computer. In 1936 *Alan Turing* wrote an article on this finding [3]. The consequences of Turing's work are an entire class of problems that are known to be *noncomputable* (i.e., impossible for computers to solve). Essentially, the halting problem leads to a conclusion that it is impossible to design an algorithm that can determine any one of many particular properties for an arbitrary program. That is we can never write a program that for every other program will determine whether the other program terminates, or calculates some particular result, or gets the correct answer.

This limitation might seem like it makes it impossible for algorithms to analyze programs in any way. However, it is important to remember that HaltChecker must work for every possible input program. It is, however, possible to write a HaltChecker that works for every program that halts. All that is needed in this case is for HaltChecker to execute that program and wait for it to complete. If the program completes, then HaltChecker can be certain that the program halts. Still, no such program works in the case that InterestingProgram fails to halt.

10.7 METAPHYSICAL LIMITATIONS

Almost from the beginning computers have been designed to mimic humans. We perceive our brains as performing two basic functions: brains remember and reason. So two of the key components of our computer hardware are the memory for remembering and the processor for reasoning. More recently computer scientists have designed software for video cameras so that they see like us, software to make robot limbs walk like us, and systems that even make the computer seem as though it is speaking like us.

If we believe science fiction writers, we might conclude that computers will replace all human function in the not-so-distant future. But what is the truth about the differences between computer systems and humans?

One of the basic premises of this book is that computer software is designed to solve problems. Solving problems requires intelligence. So it is logical to consider the question: Is a computer (program) intelligent?

The question of computer intelligence was of early importance to Alan Turing. (This is the same Alan Turing who posed the halting problem discussed in Section 10.6.) In 1950 Turing wrote an article regarding a method for determining whether a computer program could be thought of as intelligent [4].* The concept that Turing proposed has become well known as the *Turing test*. The goal of the Turing test is determine whether a computing system is intelligent.

The Turing test begins with an interrogator, a human, and a computer. The computer and the human are labeled A and B, respectively, in Figure 10.17. The interrogator cannot see or otherwise know in advance which of the two (A or B) is a computer and which is a human. The interrogator is allowed to ask questions that A and B must answer. Following numerous questions, if the interrogator cannot reliably pick which of the two is a human, then the computer system is said to have passed Turing's test. In order to maintain fairness, the questions used to be typed and answers were returned in printed form, although later versions of the Turing test also incorporate speech recognition and generation.

Turing predicted that computer systems would be able to pass the Turing test in roughly 50 years, or about the year 2000. In 1991 an annual

---

* Technically, Turing was concerned with whether computers think but acknowledged that doing well on his so-called imitation game (later renamed the Turing test) was not quite the same as evidence of thinking.

FIGURE 10.17   The Turing test.

competition, known as the Loebner Prize, was organized to see if any computer system could pass Turing's test. To date none have succeeded.

From the day it was proposed, it was clear that passing the Turing test, while an interesting measure, is not precisely the same as intelligence. Figure 10.18 illustrates some of the reasons in the form of a Venn diagram. In this figure the collection of all human behavior is pictured as a yellow dot. Further, the collection of all intelligent behavior is depicted as a red circle. The intersection of these two circles, pictured in orange, defines that human behavior that is intelligent. At best it is the orange region that the Turing test could hope to distinguish.

Surely there must be intelligent behavior outside the ability of humans. In Figure 10.18 this is pictured as the red circle that lies outside the orange intersection. The Turing test cannot hope to check for such nonhuman intelligence, because it compares a computer system to a human.

Furthermore, not all human activity can be characterized as intelligent. The nonintelligent behavior of humans is shown in the figure as the part of the yellow circle outside of the orange intersection. If, during the course of a Turing test, the human decides to behave unintelligently, then the interrogator would be trying to see if the computer system is also unintelligent.

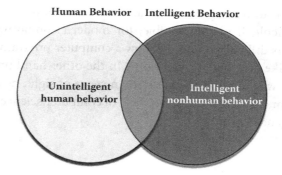

FIGURE 10.18   A Venn diagram of human behavior and intelligence.

Despite the shortcomings of the Turing test, it continues to be a respected goal on the path to creating software that is more humanlike.

One interesting outgrowth of the Turing test is the concept of a CAPTCHA. A *CAPTCHA* (short for *C*ompletely *A*utomated *P*ublic *T*uring test to tell *C*omputers and *H*umans *A*part) is a graphical image designed in such a way to make it relatively easy for a human to interpret but relatively difficult for a computer program to interpret.

Figure 10.19 contains a sample CAPTCHA. Typical of many CAPTCHAs, this is a sequence of symbols that have been distorted and perhaps colored or with extra image distractors added. Humans can easily make out the Figure 10.19 CAPTCHA as the word *inherit*, but a computer program would be unlikely to make the same interpretation in a short period of time.

The primary purpose of CAPTCHAs has been to include them on a website to ensure that it is communicating with a human. Perhaps your website wishes to administer an online survey, but needs to be certain that the survey is being completed by actual people and not some kind of automated program response. In such a situation using a CAPTCHA can be an effective way to be confident that the respondent is human. Notice how this is a kind of reverse Turing test, because this test is trying not to find the computer and human to be indistinguishable, but rather to use the differences to distinguish between human and computer.

FIGURE 10.19   CAPTCHA for the word "inherit".

In addition to intelligence there are also other human characteristics that are difficult, if not impossible, for modern computing systems to duplicate. It is difficult to imagine how a computer program could duplicate things like creativity or emotion. On the other hand progress is certainly under way to duplicate the human sense of sight, hearing, speech, touch, and smell. There is even a subfield of computer science named "artificial intelligence."

10.8 WHEN WILL YOU EVER USE THIS STUFF?

Modern media, be it popular writing, movies, or television, often leaves the impression that almost nothing is beyond the capacity of computers. Although most of this book is devoted to examining the immense capabilities of the modern computer, this chapter reminds us of the limits of those capabilities.

There are algorithms, in fact there are many algorithms, that cannot be executed quickly enough to be practically processed on a computer. Some of these algorithms, specifically those with exponential growth, will never be practical for computer processing, except in restricted cases.

Furthermore, we have seen that there are other algorithms that are inherently incomputable. The halting problem class of algorithms is impossible by their very nature and not by any limit in computer capacity.

Finally, this chapter briefly points out the challenges of creating computer systems that can exhibit things such as intelligence or emotion. Science fiction can be extremely entertaining, but it is often wise to remember that it is, after all, fiction.

REFERENCES

1. *2001: A Space Odyssey*. Written, produced, and directed by Stanley Kubrick. 1968. Warner Bros. Pictures. Film.
2. Lloyd, Seth. "Ultimate Limits to Computation." *Nature*, August 31, 2000.
3. Turing, Alan. "On Computable Numbers, with an Application to the Entscheidungsproblem." *Proceedings of the London Mathematical Society*, Series 2, 42 (1936): 230–265.
4. Turing, Alan. "Computing Machinery and Intelligence." *Mind* 49 (1950): 433–460.

TERMINOLOGY

binary search	MFLOPS
biological computers	miniaturization
CAPTCHA	MIPS
clock rate	Moore's law
duo-core	optical computers
exponential algorithm	polynomial performance
exponential growth	quad-core
gigahertz	quantum computers
halting problem	time and space
linear algorithm	Turing test
linear search	

EXERCISES

1. Suppose you work for a wireless phone provider and are placed in charge of selecting a computer program to post usage and fees for all of the customers. There are three companies that already have such programs and each offers to allow you to benchmark their programs. Explain things you could do in designing benchmarks to help to select a program.

2. Explain some of the potential things that would make the benchmarks in Exercise 1 less valuable.

3. Classify each of the following algorithms as linear, polynomial (but slower than linear), or exponential. Consider the algorithmic growth relative to N. That is, as N grows how does the algorithm grow?

   a. Surveying N people in a room to find out the political affiliation of each

   b. Recording each score for a round robin tournament for N teams (in a round robin tournament each team must play against every other team)

   c. Counting how many credit cards have a negative balance from a collection of N credit cards

   d. Filling in each two-dimensional table that shows the distance between each pair of airports for a total of N airports

   e. Listing all possible automobile license plates of length N letters

4. What is the exact number of different combinations for a padlock, like the one described in Section 10.5, but having only 20 different numbers around the dial?

5. Card counters are often barred from casinos because they are able to alter the odds of a card game by their unique ability to remember which playing cards have been dealt. This is helpful in games such as blackjack where playing cards are discarded after they are dealt, so the card counter knows what cards remain in the dealer's deck to be dealt in a future game.

   a. Suppose the dealer's deck contains 52 cards. How many different two-card hands can you be dealt?

   b. Suppose the dealer's hand contains only 10 cards. How many different possible two-card hands can you be dealt?

   c. The algorithm used by card counters is only useful when the remaining number of cards becomes quite small. What kind of algorithmic growth is being exhibited by the number of potential hands?

   d. Suggest a way that the casino could eliminate the problem with card counters.

6. Consider the Turing test.

   a. Suppose your friend writes a computer program that passes the Turing test. Explain why this would be significant.

   b. In what ways is passing the Turing test different from intelligence?

7. Which of the following is the best way to explain the significance of the halting problem?

   a. The halting problem shows that it is possible to solve any problem with a computer program.

b. The halting problem shows that determining whether a computer program halts can be translated into a computer program.

c. The halting problem demonstrates that some algorithms cannot be solved by today's computers, but may be solved by faster computers in the future.

d. The halting problem results in a proof that there are some problems that can never be solved by a computer program.

8. Postal services commonly use computer programs to analyze written addresses on the outside of letters and parcels.

a. What concept discussed in this chapter becomes less useful as such address recognition programs improve?

b. What do you think could be done to improve the reliability of CAPTCHAs?

9. Identify each of the following algorithms as possible (this algorithm could be executed on a computer), noncomputable (this algorithm can never be written or executed by any computer), or impractical (this algorithm has exponential run time and requires too long for computers to execute, except in restricted cases).

a. A program to crack any decryption scheme a terrorist could use.

b. A program to analyze any other program and tell you if it solved your particular problem.

c. A program to calculate the census for the United States.

d. A program to analyze traffic flow through a city by examining every possible combination of possible turns for every automobile in the city.

e. A program to print all of the words in an English dictionary.

f. A program to explore every possible search engine expression compared to a single web page. (You may assume that the search expression is limited to say 256 characters in length.)

CHAPTER 11

Concurrent Activity

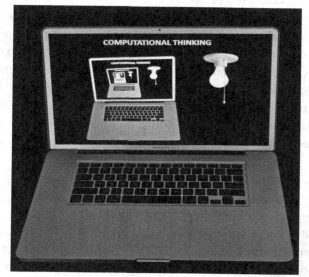

Multitasking? I can't do two things at once. I can't even do one thing at a time.

—HELENA BONHAM CARTER

OBJECTIVES
- To be able to explain the difference between parallelism and concurrency, and the role of supercomputers and distributed computing in modern problem solving
- To recognize basic constraints that prohibit simultaneous execution

321

- To be able to interpret timing diagrams that illustrate concurrent behavior
- To be able to trace the behavior of a given sorting network for a specific data set
- To be able to explain the potential of concurrency for improving performance
- To be able to explain how shared resources can constrain concurrency
- To recognize a TOCTOU (race condition) and be able to explain how these situations can lead to errors
- To recognize deadlock and live lock situations

Events do not often occur in a nice, neat, one-at-a-time sequence in real life. Sometimes the wind blows at the same time the rain falls. Other times you might choose to walk and talk simultaneously. Aircraft often bank and accelerate all at once. Not surprisingly, computer programs can also be expected to multitask as well. When software can perform multiple tasks at once, we call it *concurrent execution*, or just *concurrency*.

The discussion of concurrency has been delayed until this point, because humans are more adept at a single line of reasoning. As a result, thinking in nonconcurrent ways and writing nonconcurrent software tends to be easier than their concurrent counterparts. This chapter explores some of these challenges, along with both the advantages and the pitfalls of concurrency.

11.1 PARALLELISM OR CONCURRENCY?

As computer hardware marches faster and faster, a significant performance barrier looms. Electricity travels by moving electrons at roughly the speed of light. As mentioned in the previous chapter, miniaturization has led to circuits in which the width of electrical pathways can be measured in molecules. Reducing circuit sizes (and, therefore, increasing their performance and storage capacity) is becoming increasingly expensive. At some future date such miniaturization will no longer be cost effective.

An alternative to miniaturization becoming popular among computer processor manufacturers is to include several processors in the same integrated circuit. Sometimes, these processors are referred to as *cores* leading to the name *multicore processors*. The basic idea is that each processor (or core) can execute its own instructions at the same time that other processors are executing other instructions. Such simultaneous execution by multiple devices is known as *parallel execution* because the devices are executing in parallel.

The ultimate in parallel execution are also the fastest computers of our day, known as *supercomputers*. The speed of these machines is measured in *petaflops*, quadrillions of instructions executed per second. These supercomputers are something like 100,000 times faster than your personal computer. They accomplish such feats not by the speed of their individual processors but rather by using millions of processors executing simultaneously.

A different kind of parallel execution, known as *distributed computing*, occurs every day via the Internet. Distributed computing occurs whenever multiple largely independent computers work collectively to solve a common problem. The Internet itself is an example of distributed computing, because all computers connected to the Internet participate in the task of communication with one another. More recently known as *grid computing*, distributed computing has been used to solve such problems as weather forecasting and financial modeling. In addition the term *cloud computing* has also come to refer to a type of distributed computing that involves significant amounts of shared data.

Folding@home is an example of one such distributed application. Begun in 2000 at Stanford University [1], this project has enlisted the voluntary participation of hundreds of thousands of Internet computers. The program is designed to utilize otherwise unused processing capacity of your computer. The collective problem solving involves the study of protein folding (an area of research in chemistry) and how such molecular deformation impacts different diseases, such as Alzheimer's, Parkinson's, Huntington's, and various forms of cancer.

Another example of grid computing is *climateprediction.net* or CPDN. Tens of thousands of computers in over 100 countries volunteer their computers part time to participate in this Oxford University research. The purpose of CPDN is to model global climate changes. One early result from CPDN was described by Piani et al. [2].

It is generally more proper to use the term *parallel computing* when discussing multicore hardware, supercomputers, and distributed computing. However, the distinction between parallel computing and concurrency is largely a technical detail for our purposes in this book. We choose to use *concurrency* for all forms of simultaneous or potentially simultaneous execution in our further discussion because it is typically associated with software systems, whereas *parallelism* is more appropriate for describing hardware forms of simultaneous execution. Further, concurrent execution subsumes parallelism so we make no further distinction.

FIGURE 11.1   Tournament brackets for four-team single elimination.

11.2 SCHEDULING

To understand the concept of concurrency, consider a single-elimination soccer tournament. (In single-elimination tournaments each team continues to play other teams until their first loss, with the final unbeaten team declared tournament winner.) Suppose that there are four soccer teams—named Jives, Tangos, Rumbas, and Polkas—in a tournament. Figure 11.1 is a bracket diagram often used to picture such a tournament. These brackets show that three games are required: a game between Jives and Tangos, a game between Rumbas and Polkas, and a game between the winners of the first two games.

The Figure 11.1 bracket diagram depicts which games are played, but not when or where. Suppose that we only have a single soccer field on which to play the tournament's games. In this case only one game can take place at a time. This means that each game must be scheduled for play at a time different from every other game. Assuming that each game takes three hours, the tournament can be described by the timing diagram of Figure 11.2.

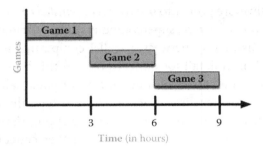

FIGURE 11.2   Timing diagram for four-team tournament played on one field.

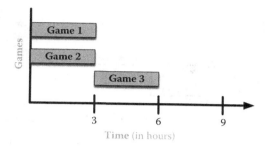

FIGURE 11.3   Timing diagram for four-team tournament played on two fields.

The timing diagram uses a rectangle to represent each soccer game. The left edge of the rectangle shows the time when the game begins and the right edge is positioned at the ending time. If the three games are played consecutively, then the entire tournament can be completed in nine hours, as shown.

Our tournament can be thought of as an analogy to tasks executing on computer processors. The play of each tournament game is analogous the execution of a single software task and the soccer field is the processor upon which the task executes. Playing the entire tournament on a single field is like having but one processor to execute all the tasks. In this case a minimum of nine hours is needed to execute these three tasks.

Scheduling consists of determining which tasks to execute on which processors at what time. Scheduling becomes more complicated with multiple processors. Suppose the soccer tournament is played on two fields (analogous to two processors). In this case it is possible to play games 1 and 2 simultaneously, each on a different field. Figure 11.3 shows the timing diagram for such a two-field tournament.

The obvious advantage of multitasking is that it can save time. The soccer tournament using two fields is played in two-thirds the time of a one-field tournament. But more processors do not always improve performance. A four-team tournament would not benefit from having a third field, because game 3 cannot be played until the winners of games 1 and 2 have been determined. Such dependencies between processes must be taken into account when tasks are scheduled.

Suppose instead of four teams, there are eight teams in the tournament. Figure 11.4 shows the brackets for an eight-team, single-elimination tournament.

If four or more soccer fields are available for the tournament, then games 1 through 4 can all begin at once. None of the other games can be

FIGURE 11.4 Tournament brackets for eight-team single elimination.

played at beginning of the tournament, because the other three games all depend upon the outcomes of some of the first four games. Further note that games 5 and 6 can be played simultaneously, but game 7 is required to follow their completion. Figure 11.5 is a timing diagram to show the most efficient way to conduct an eight-team tournament on four fields. Note that more than four fields would not shorten the total tournament time.

The eight-team tournament demonstrates the kind of performance improvement potential of concurrency. The seven-game tournament is completed in nine hours, while the same tournament requires

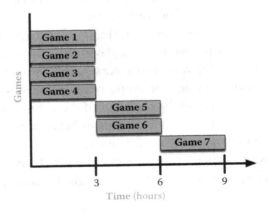

FIGURE 11.5 Timing diagram for eight-team tournament played on four fields.

twenty-one (7 games times 3 hours per game) if all games are played on the same field. This is like a concurrent computation for seven tasks executed on four processors. The tournament is completed 2⅓ times faster using multitasking.

11.3 SORTING NETWORKS

To further explore the utility of concurrency this section examines a common computer problem—data sorting—using both concurrent and nonconcurrent solutions. *Sorting algorithms* reorder a collection of data so that it is arranged in either ascending (smallest to largest) or descending (largest to smallest) order.

Sorting clearly did not begin with the advent of computers. Maybe even cavemen arranged their stones from smallest to biggest (Figure 11.6)? Teachers are sorting in descending order whenever they arrange homework papers from highest score to lowest. A postal carrier sorts mail for a single street in order to match the ascending or descending order of street addresses. Alphabetization is also sorting, but sorting by the order of letters rather than numeric order.

If asked to sort a few telephone numbers in ascending order, you could do so fairly quickly. But sorting hundreds or even thousands of numbers is a time-consuming and tedious undertaking. But the time-consuming tedium is what makes the computer an excellent candidate for sorting. Computers sort students by ID number. They sort your income tax by

FIGURE 11.6 Caveman with sorting rocks.

Social Security number. They sort driver's license numbers, checking account numbers, user IDs, e-mail addresses, and so on.

Sometimes you can choose different ways to sort the same data. For example, an online auction might allow you to sort merchandise from least expensive to most expensive or the opposite. You might also be able to sort by distance from your home or by the ending date of the auction or even by the date that this item was first listed.

Have you ever thought about the algorithm that you used to sort? Computer scientists do. In fact there are many algorithms for sorting, and most proper sorting algorithms work well for small amounts of data. You may recall that Chapter 10 includes one example of a sorting algorithm suitable for use on a single processor. However, sorting large amounts of data can take time. So sorting is a good candidate for the use of concurrency.

Imagine that you were asked to sort a room full of paper files, but your manager realizes that the job is too big for one person. So you are assigned a team of people to help. Like any concurrent procedure, the first step is to determine how to use the extra people (processors) to make them productive.

One way to sort concurrently is to use a *sorting network*. You can think of a sorting network as a collection of identical simple processors. A single processor can (1) receive two values from elsewhere in the network, (2) compare the two values, and (3) order the two compared values by transmitting them appropriately in the network. Figure 11.7a pictures a single such processor.

The four-sided figured in the center of Figure 11.7a represents the processor. Think of the arrows as data transmission lines that send or receive data, one value at a time, in the indicated direction. The two arrows on the left are incoming lines from other parts of the network. The outgoing (rightmost) arrows are used by the processor to transmit its data elsewhere. The work of the processor is to receive a pair of values from the incoming lines, then transmit these same values so that the smaller of the two is sent on the top (<) line and the larger is sent on the bottom (>) line. (If the two values are equal, then the same value is sent simultaneously on top and bottom outgoing lines.)

Suppose that the processor receives the integer 5 on the top incoming line and the integer 3 on the bottom. In this case the processor would conclude that 3 is less than 5, so 3 is transmitted out on the top line and 5 on the bottom. If the processor were to receive 3 on the top incoming line and 5 on the bottom, then the same thing happens to the outgoing lines.

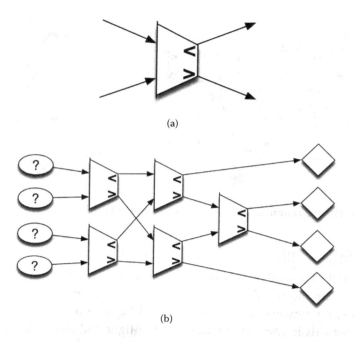

FIGURE 11.7 (a) A single sorting network processor. (b) A sorting network consisting of five processors.

A single processor like that shown in Figure 11.7a can sort two pieces of data. Sorting more data can be done with more processors, but they must be arranged and connected properly. Figure 11.7b shows a sorting network with five processors that is designed to sort four values. The ovals on the left of this diagram depict the source of data; each question mark would be replaced by one value for sorting. The diamonds on the right side of the figure represent containers for the final data. If we connected our processors correctly, then after an appropriate length of time the data will be sorted in ascending order top to bottom within the diamonds.

Figure 11.8 shows the same 4-input sorting network that has sample input. In this case the four values to be sorted are the integers 4, 3, 2, and 1. The behavior of the network for this data is shown by labels on the network arrows. Each arrow is labeled to show the value that is transmitted over this line. For example, the upper-left processor receives 4 and 3 on its incoming lines and sends the 3 out on its top line and the 4 on its bottom line. The upper-right processor receives 3 and 1, sending the 1 out on its top line and 3 on the bottom. The values received by the diamonds on the right are sorted.

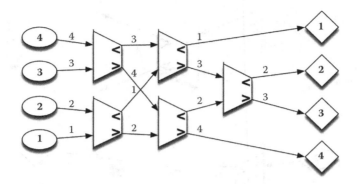

FIGURE 11.8   Example of the four-value sorting network.

11.4 MEASURING CONCURRENCY'S EFFECT

The speed of a sorting network, like most other computation, is mostly governed by the speed of the processors. The only other potential delay in a sorting network is the time required for data to travel through the lines, but this is governed by the speed of light and should be negligible by comparison to the processing time. Therefore, to analyze the speed of a sorting network we shall assume that the processor requires one unit of time in order to compare its two incoming values and send them properly to the outgoing lines.

The five processors shown in Figure 11.8 must each perform one comparison, and therefore consume one unit of time. However, some of the processors can work simultaneously. Let's assume that all incoming data is available at the same time, as shown in the Time 0 portion of Figure 11.9.

The leftmost two processors can perform their work at the same time; their results are depicted in the Time 1 image. The center two processors cannot compare data, until the leftmost processors have completed their work by transmitting the reordered results; this happens at Time 2. Following Time 2 the data is available for the rightmost processor to complete its work in Time 3.

Figure 11.9 shows that three time units are needed to complete the sort. These same three time units would be needed regardless of the data to be sorted. The factor that limits concurrency in sorting networks is the arrival of data to a processor when the data comes from other processors.

Figure 11.10 shows a sorting network for sorting eight values. This network uses a similar pattern as the four-value sorting network. In each case the network begins with a matrix of processors in which the number of

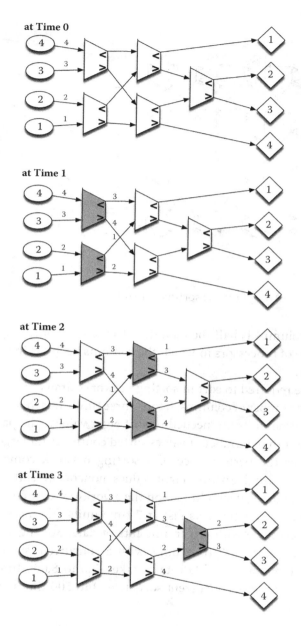

FIGURE 11.9 Timing of a sort in a four-value sorting network.

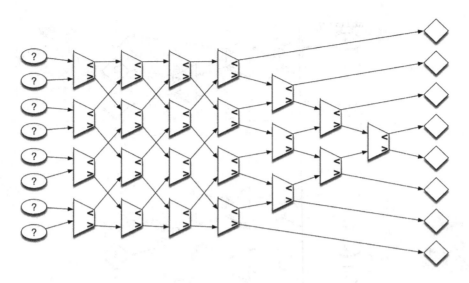

FIGURE 11.10   An eight-value sorting network.

rows and columns is half the data size. To the right of the square matrix is a triangle of processors in which the columns decrease in height by one left to right.

The time required to sort in sorting networks drawn in this way is the same as the number of columns. It took three time units to sort four values, and seven time units are needed to sort eight values. The pattern of one less time unit than number of values sorted continues for larger networks.

How does the performance of a sorting network compare to non-concurrent sorting? Sorting eight values nonconcurrently takes about 15 to 30 time units, depending upon the algorithm. Our concurrent sorting network is over twice as fast at 7 time units. The performance differences accentuate with larger amounts of data. Sorting 128 items can be accomplished in 127 time units with a concurrent sorting network. Nonconcurrent sorting of 128 items takes 700 to 8,000 time units. For 1,000 data times, nonconcurrent sorts take 9 to 500 times longer than a sorting network.

11.5 CHALLENGES OF CONCURRENCY

It might seem that 100 processors should be able to complete a task 100 times faster than a single processor. However, this is rarely the case. This tends to be born out by human activity. If we think about people as processors, then a human analogy might be to use 100 people working collectively on the same task. Rarely would we expect 100 people to complete a task 100 times faster than a single person. With additional people comes the need for additional communication and additional coordination, both of which take time. Also, care must be taken to ensure that the work of one person does not interfere or duplicate that of another, and there is the basic problem of determining how to assign tasks in a way to keep everyone productively busy.

The single elimination soccer tournament and sorting network examples illustrate how the number of concurrent processors is generally more than the increase in speed. Furthermore, these two examples of concurrency point out some of the specific constraints that limit concurrency performance. For both concurrent soccer games and sorting networks there are shared resources that restrict when processors may execute. The sorting processors cannot begin to execute until both input data values are available, but these values are a kind of shared resource with other processors, at least the processors that previously compared the values.

Recall that in the soccer tournament example it is the soccer field that plays the role of processor, and a processor execution consists of playing games. The constraint upon beginning a new game is that the teams are available, and this often means that teams have won a previous game. Therefore, the shared resource in this case is the soccer team. A team must finish a prior game on either the same or some other field before the next game can start.

Failing to observe concurrency constraints causes multitasking to fail. If a processor in the sorting network is allowed to compare values before they are available, what will happen? How could the same team even play two soccer games at once?

There is more to the shared resource issue, and to understand this we turn to a third example of a group document editing system. Figure 11.11 represents a system that is fairly common in cloud computing environments. The four people in the system (Calista, Eric, Sandra, and Tim) are all able to share the same documents in a collaborative environment. One person might create a document and place it in the shared document area (depicted as a cloud). Anyone of the four people can read or edit any of the shared documents.

Perhaps this shared document system uses a separate computer for each user and a shared server to manage the documents. In this way it is easy to think of each person (more specifically, each person's computer) as a separate processor that operates concurrently with the other three.

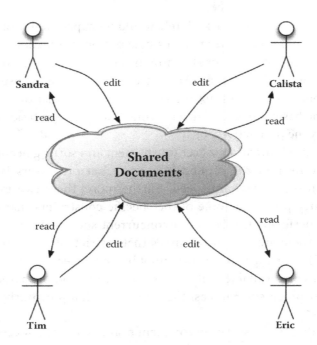

FIGURE 11.11   A four-person shared document system.

Not all shared resources lead to concurrency constraints. Suppose that Sandra and Tim wish to read the same shared document at the same time. This kind of simultaneous reading poses no problems. In fact all four users could be reading the same document at the same time without difficulty. This kind of shared read access is no different than multiple Internet users viewing the same web page at once, and it demonstrates that sharing simultaneous read access is not problematic.

However, potential problems can occur when one user is reading a document while someone else is altering the document. If Eric is reading a document while Calista is editing the same document, then which version is Eric reading: the one before Calista edits or Calista's latest version?

The problems tend to get worse when there is multiple concurrent editing (updating) of shared documents. Suppose that Sandra and Calista are both editing the same document. Both examine the document at the same time and both simultaneously decide they dislike the title. If Sandra and Calista edit the document title simultaneously, then what happens to the title? Is the new title Sandra's, Calista's, or some mash of the two? Furthermore, it might be that Sandra would have preferred Calista's title or Calista would have preferred Sandra's, but they both see only the original title before editing.

Lest this seem like an unimportant issue, consider that the shared document is a list of seats available for a concert and that two users both attempt to reserve the same available seat. Unless the software takes care to constrain such concurrent access, it is possible that both individuals can reserve the same seat. This is not a silly example, it actually happens when computer programs fail to handle concurrency properly.

The common problem in these examples is called a *time-of-check time-of-use (TOCTOU) condition*. A TOCTOU condition occurs when there is simultaneous update access to a shared resource. TOC refers to the time that a processor checks the status of a resource and TOU refers to the time that a processor updates a resource's status. Reading the document title is a TOC activity and editing the title is TOU; together they form a TOCTOU condition because another user is also checking and updating the same document. TOCTOU occurs in the concert seat example because both users check (TOC) for a seat at the same time, before reserving (TOU) the seat.

The TOCTOU situation that needs to be avoided occurs when one process performs a check on a shared resource, while the second process performs both a check and update on the same resource, before the first process has time to complete its update. For example, Tim checks that a document exists, then Eric discovers the document exists and deletes it before Tim tries to edit the same document. What happens when Tim tries to edit the nonexistent file? Sometimes TOCTOU conditions are called *race conditions* because the most problematic situations occur when one process races to perform both a check and an update between another check and update of the same resource.

Race conditions occur in real life as well. Suppose you have slow reflexes while approaching an intersection in an automobile. You check for opposing traffic to your right and discover there is none. However, before you enter the intersection another car races in from the right, checks to find no traffic to its right and enters the intersection. Having once checked to your right, you enter the intersection, and both cars collide (Figure 11.12).

TOCOU conditions are problematic when the check and update of a given resource are separated by another racing processor. This problem can be eliminated by prohibiting the separation of the check from the update. Instead, it is better to make the check plus update a single indivisible operation. Once one processor initiates a check, it is said to *lock* the

Step 1
Coming to an intersection, you push in the
clutch and check for traffic to your right.

Step 2

You leisurely put the car into gear...
Meanwhile, a car comes racing from your right, checks to its left and
right and enters the intersection.

Step 3
You let out the clutch, lurch forward and CRASH!

FIGURE 11.12   Two automobiles experience a race condition.

resource, and the resource is not unlocked until its update is completed.
A locked resource cannot be accessed in any way by other processors,
forcing such processors to wait until it is unlocked. Only the process that
performed the lock is allowed to unlock, giving the process ownership of
the lock. When checking for an available theater seat, that seat becomes
locked by the person checking; this seat is unavailable to others until the
checking individual has an opportunity to reserve the seat.

Automobile drivers often employ this kind of lock solution without
knowing it. For example, when approaching a four-way stop intersection
the unwritten rule is first come, first served. In effect the first car to come
to a stop has a lock on the intersection. Locks are also used for upgrading
web browser programs. An executing web browser has a lock on its own
program so that even an attempt to update the program cannot occur until
the web browser quits executing thereby relinquishing the lock. Without

such a lock, the web browser's execution and the administration of an update would create a race condition on the stored web browser program.

A locking mechanism eliminates TOCTOU condition problems, but it does pose constraints on concurrency. If five people try simultaneously to place a bid on an online auction item, then one of them will hold a lock on the item, and the others cannot place their bids until the first person is finished and the lock relinquished. The lock effectively prohibits some concurrency.

A more severe difficulty posed by locking mechanisms is known as *deadlock*. Deadlock occurs whenever processors have locked resources in such a way that they cannot proceed. For example, suppose that Kelly and Derek each want to arrange separate three-way conference calls with Kasandra and Mike (see Figure 11.13). At step 2 Kelly connects to Mike; meanwhile Derek connects to Kasandra. When in step 3 Kelly and Derek try to call the remaining person, they are unable to do so, because they are already committed to a different conference call; effectively their phones are already locked and cannot accept a different call.

Deadlock occurs when processors have locked resources in such a way that the processors cannot continue to execute without access to resources locked by other processors. In this case Kasandra's and Mike's phones are resources with Mike locked by Kelly and Kasandra locked by Derek. Each of these phones is effectively locked so neither three-way conference call can be complete until either Kelly or Derek decides to end their call (i.e., relinquish their lock).

The term *gridlock* is commonly used when deadlock occurs in traffic. Figure 11.14 depicts an example of gridlock. Think of the six cars as processors and street segments as resources. A car has a lock on a segment of the street when it occupies that segment. In the picture each car occupies a part of the road that is needed by another car in order to drive forward. The six cars are deadlocked with no car able to move forward.

True deadlock occurs when it is impossible for processors to execute without one or more relinquishing resources. The effect is that none of the processors, like the cars in Figure 11.14, can make progress. Sometimes concurrency does not lead to true deadlock, but it still seems that way to some of the processors. For example, a timid driver trying to enter a on-ramp of a congested freeway may find it impossible to advance even though there is adequate space for a more aggressive driver. The timid driver would feel the same effect as deadlock, even though there is no deadlock. This situation is called *live lock*, because the processors are not deadlocked, but one (or perhaps more) processor is effectively blocked permanently.

Step 1
Kelly wants to place a conference call to both Mike and Kasandra.
Derek also wants to place a conference call to Mike and Kasandra.

Step 2
Kelly calls Mike who answers.
Derek calls Kasandra who answers about the same time.

Step 3
Kelly tries to call Kasandra and Derek tries to call Mike.
DEADLOCKED — unable to complete conference calls.

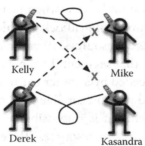

FIGURE 11.13   A deadlock from attempting concurrent conference calls.

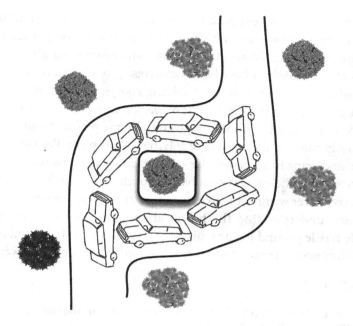

FIGURE 11.14   Six cars in gridlock.

Live lock occurs in computer systems when processors are not treated fairly in granting resources. If the system's resources are quite busy, sometimes a particular processor or a particular task is assigned such a low priority that it never makes any progress.

11.6 WHEN WILL YOU EVER USE THIS STUFF?

Google the phrases "distributed computing" and "multiprocessor computers" and you will discover two of the most important advances of the past decade. The ubiquity of the Internet and of cell phone networks open countless possible ways to apply collections of interconnected computers for solving a common problem. Consider using a GPS app on your smartphone as you drive around Chicago. A cleverly designed app might take advantage of concurrency to collect information from you and other users in the same vicinity in order to estimate how traffic is flowing on specific Chicago streets. This information is reported back to be displayed on your GPS as advice on which streets provide the fastest route to your chosen destination.

The same kind of concurrent thinking that lead to using a GPS in this way could also lead to more success in playing any game that involves multiple people making concurrent decisions. Such a multiuser game is not so different from a business world in which each corporation strives to succeed in an environment where all competitors change and adapt in parallel.

Recently, researchers have begun to investigate using a concept known as *crowd sourcing* as a means for solving complex problems. In a crowd sourcing solution, humans play the role of concurrent processors. All humans in the group (crowd) are given the same problem and are able to collectively share in the process of solving the problem. Perhaps someone notices and error in someone else's approach, or perhaps one person can suggest a more efficient way to solve part of the problem. The point is that even if you never write a concurrent computer program or operate a supercomputer, understanding the potential and limitations of concurrency provide fertile ground for new discovery. Crowd sourcing is but one way in which concepts from concurrency are used in nontechnical ways.

REFERENCES

1. *Science Daily,* 2002. http://www.sciencedaily.com/releases/2002/10/021022 070813. htm. October 22 (Accessed Dec. 5, 2013).
2. Piani, C., Frame, D. J., Stainforth, D. A., and Allen, M. R. "Constraints on Climate Change from a Multi-Thousand Member Ensemble of Simulations." *Geophysical Research Letters* 32(23) (2005). L23825

TERMINOLOGY

climateprediction.net (CPDN)	multicore processor
cloud computing	parallel execution
concurrency	petaflop
core	race condition
deadlock	scheduling
distributed computing	sorting algorithm
folding@home	sorting network
grid computing	supercomputer
live lock	TOCTOU condition
lock (of a resource)	

EXERCISES

1. Figure 11.4 shows the pairings for a single-elimination tournament for eight teams. Consider how this might change if twenty-four teams rather than eight were playing.

   a. Draw a diagram for the tournament.

   b. How many fields would be required to maximize concurrency? (Please assume that all games require the same length of time to complete and remember that a game for one round cannot begin until the winners of the previous round are known.)

   c. Draw a timing chart, similar to the one in Figure 11.3, to describe the 24-team tournament.

   d. How much of a performance gain (how many times faster) is this tournament than if it were played on a single soccer field?

2. The single-elimination tournament example is very much like a searching algorithm that searches for the maximum of a collection of values. (Think of each match of the tournament as one comparison with the winner being the maximum value of the comparison.)

   a. If you were to calculate that performance of the algorithm for calculating a maximum by using one time unit per comparison and ignoring all other time, then the nonconcurrent algorithm would require 99 comparisons to find the maximum of 100 numbers. How long would a concurrent algorithm require for the same 100 numbers? (Note that you should allow for comparisons to be concurrent using a model similar to a single-elimination tournament.)

   b. Explain a way to use processors so that you could find the minimum number in no more time than it takes to find the maximum, assuming the performance described in part (a).

3. Three families (Riley, Hunt, and Singh) each want to purchase a new home. Since each family already owns a home, they would like to sell their home before committing to a new purchase. Realtors support this kind of activity in the form of a contingency offer. Such an offer is a legal agreement in which one family agrees to purchase a new home only if their own home is sold first. Explain how the three families could end up in deadlock by using contingency offers. How

could the deadlock be resolved, once deadlock has occurred? What policy would have been prohibited in the first place?

4. Trace the behavior of the sorting network from Figure 11.10 assuming the following order of values from top to bottom: 8, 3, 1, 7, 2, 5, 4, 6. Be sure to annotate each arrow with the value that flows in this direction.

5. Redraw the sorting network from Figure 11.10 so that it works for only six values.

6. Following is a proposed sorting network for six values with a different geometry than those shown in the chapter. Does this network sort properly?

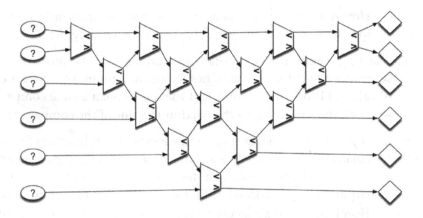

CHAPTER 12

Information Security

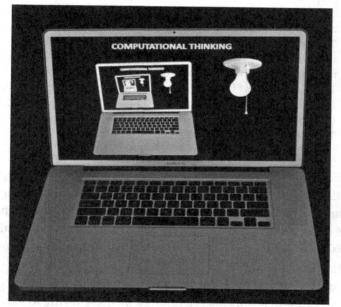

Better be despised for too anxious apprehensions, than ruined by too confident security.

—**EDMUND BURKE**

OBJECTIVES
- To be familiar with basic security-related vocabulary
- To understand that security has three ingredients: confidentiality, integrity, and availability

- To know common forms of cybercrime, including viruses, malware, spoofing, relay attacks, network sniffing, spamming, shoulder surfing, dumpster diving, identity theft, denial of service, social engineering, and phishing
- To understand commonly used authentication technologies and how they work
- To recognize two-factor authentication and how it provides better security
- To recognize the difference between read, write, execute, and own authorization
- To understand the need for uniqueness in IDs that are used for security purposes
- To be able to explain risk as a two-part ranking of likelihood of breach and potential cost of the damage
- To understand the purpose and time to use common mitigation strategies, including one-way and two-way encryption, firewalls, antivirus software, software update, file backup, and log files
- To recognize and be able to apply a few basic security principles, including to secure the weakest link, reduce the attack surface, defend deeply, compartmentalize, and trust reluctantly
- To understand how openness contributes to security

From the very first chapter this book has been about how computer scientists think. Unfortunately, a few computer scientists use their skills to do harm. The problem is that just as computing technology can be used to solve our problems, so too it can be used to steal or vandalize. It is this possibility for abuse of technology that has elevated the importance of computer-related security.

The topic of security is in reality a modern struggle between good and evil. Computational thinking is employed by those trying to attack others via the Internet. Computational thinking is also used to construct security systems that counter such threats, and all of us need to think computationally for our own protection in this age of technology.

12.1 WHAT IS SECURITY?

In the agrarian community of our ancestors security was probably about guarding crops from vandalism or theft. Following the Industrial Revolution manufacturers worried about securing not only the machines, but also the plans and ideas for future inventions. The Age of Information, as many people call our modern world, has found ways to store not only ideas and plans, but also our messages, blogs, and tweets. So it is not surprising that the topic of *information security* is of paramount importance. But what exactly do we mean today when we talk about information security or computer security?

Before answering the last question, we need to review some basic terminology. Our vocabulary begins with the notion of an *asset*, which is the object that needs protection. The person that could potentially do harm to our asset is known as an *attacker*. The ways in which the asset is susceptible to attack are call *vulnerabilities*, and any attempt to compensate or diminish vulnerabilities is known as *mitigation*. *Security systems* are procedures, policies, devices, and mechanisms designed explicitly to mitigate. The goal of an attacker is to *exploit* a vulnerability by treating the asset in some undesirable manner. Any possible exploit poses a *threat* to the asset, and a successful exploit (one that does actual damage) is known as a breach.

To illustrate these concepts consider the days of the cowboys, shortly after the western United States had been settled. The main asset of the cowboy was the cattle he guarded. Cowboys referred to attackers as "cattle rustlers." Cows were vulnerable to theft, injury, and murder; the greatest threat was probably theft because injuring an animal was of no practical value to a rustler. Cowboys had various ways to mitigate the threat of theft, including building fences and branding animals with symbols that proved ownership. The cowboys, themselves, were also part of the security system as they patrolled around cowherds on horseback. Unfortunately, rustlers sometimes still found ways to exploit vulnerabilities in the security system and abscond with free beef.

The cowboy example may be useful for exemplifying vocabulary, but in this book we are more interested in security as it applies to information stored in computers. In this context the vulnerabilities are far more varied and complex. However, some of the same threats of concern to cowboys are still applicable.

The term *physical security* refers to securing the hardware devices, because an attack could simply be to steal your computer. Since information is stored in specific devices, the greatest physical security assets are probably hard disks, flash memory sticks, CDs, and DVDs. Exploits of physical security typically require very little technical sophistication on the part of the attacker. For example, passwords have been stolen in all of the following ways: (1) reading a password from a piece of paper in the victim's office, (2) observing the victim log in, (3) discovering a password within the victim's trash, and (4) gleaning a password from a file the victim created on a computer that was logged in and unattended.

The 2010 E-Crime Watch Survey collected the responses of 531 security professionals from the United States [1]. Some of the findings of this survey include the following:

1. Sixty percent of the respondents had experienced at least one cyber security event in the past year. The mean number of such events was five per respondent.
2. The total monetary loss due to cybercrime was more than $200 million.

Good problem solving searches for the root cause of any difficulty, and information insecurity can be traced back to one of three sources:

1. Hardware failure

2. Software failure

3. Human failure

The cowboys of yesteryear had to deal with hardware failure whenever a saddle broke, a gun failed to shoot, or the horse bucked its rider. Cowboys also coped with human failure whenever someone forgot the location of their cattle or lost sense of direction. However, cowboys did not need to worry about vulnerabilities introduced by software. The state of the art in computer software is that for every few thousand lines of code we can expect a fault. Most of these faults do not result in significant security vulnerabilities, but some do. A user can expect both application software and operating system software to introduce a small number of threats.

Whenever a computer receives new content there is always a possibility that the content allows for exploits of one kind or another. Security professionals refer to these external delivery mechanisms as *vectors*.

Historically, motion pictures seemed to partition cowboys into two groups: good cowboys and bad cowboys. The good cowboys were typically clean-shaven and handsome, whereas the bad cowboys were more often rather scraggly and unwholesome in appearance. Another interesting characteristic of these western movies were that the bad cowboys were nearly always costumed in black hats, while the good cowboys sported white hats.

Today, we use the terms *white hat* to refer to actions that are good from a security point, and undesirable (attacker-like) actions are termed *black hat*.

Like many technologies, vectors, can be used for either black hat or white hat purposes. A DVD can be used to install new software or to play a video; these would be considered white hat uses. However, DVDs are also effective vectors for black hat purposes, such as the delivery of pirated music or delivering a computer virus.

Actually, information security was not viewed as a serious concern until the advent of networks. There are many reasons why the Internet introduces security vulnerabilities. In addition to providing a convenient vector for attackers, the Internet connects computers with other computers and servers, each of which may have their own software flaws and their own poor physical security. In addition the communication protocols on which the Internet relies are often not designed to be particularly secure. Still another vulnerability inherent to networks is that they can be configured in ways that are exploitable. All of this means that an attacker on the other side of the world might be able to successfully steal or damage your information because of the way that the Internet was designed, or due to a flaw in some Internet server in a third country, or because some network administrator accidentally forgot to set a security restriction properly or just because you forgot to lock your office door. The point is that the options for black hat uses of technology are vast and the Internet is a contributing factor.

12.2 FOUNDATIONS

Computational thinking requires that any new problem be carefully analyzed. So we begin by considering the issue of what constitutes security. If we say that we are secure, what do we really mean?

Security experts who have studied this question, identify three characteristics of security. We can think of these as the pillars of information security. To be secure information must

1. Be *confidential*
2. Have *integrity*
3. Be *available*

In other words a complete security system must provide for these three aspects: confidentiality, integrity and availability (Figure 12.1).

In order to be confidential, access to information must be properly restricted. Only the persons, processes, or devices that are authorized are permitted access by a truly secure system.

FIGURE 12.1 The three foundational aspects of security.

To say that information has integrity means that the information is reliable and trustworthy. We might say that information has integrity when it has not been corrupted in any way. That said, integrity is somewhat more complicated by computer networks. In a world that is connected by the Internet information moves freely and it becomes important to associate responsibility. We can think of the information as having two parts: the data that is stored and the party who is responsible for that data. (By the phrase "party responsible" we are referring to a person, process, or device.) When an e-mail message is sent, it is important to know both the e-mail content and the individual who sent the e-mail. Similarly, when a packet of data is sent through the Internet, it is important to know the data contained in the packet and also the source computer that sent the packet.

This two-part nature of information results in two types of integrity—*data integrity* and *owner integrity*—and both of these are required to establish full integrity. To preserve data, integrity means that the information's content (the data) has not been altered from its correct value. We use the word *owner* to refer to the person, process or device that is responsible for the data. Information is said to have owner integrity when this ownership is correctly identified.

The topic of ownership is often not so simple. Sometimes information is anonymous; in other words, no one claims ownership. The opposite extreme is when multiple people claim ownership. Children make use of multiplicity of ownership when each claims that the other is at fault.

Without *identity* there can be no owner; and without an owner, integrity of ownership is meaningless. Therefore, security systems are dependent upon identity. Every user, every process, and every device that owns assets needs a unique identity within a security system. E-mail addresses, user account IDs, and the IP numbers used to distinguish computers connected to the Internet are all identifiers. You cannot receive e-mail without an e-mail address. You cannot log in to a network server without a user ID.

The first two things that likely come to mind when you hear the word *security* are confidentiality and integrity, but these two qualities are not enough for information to be truly secure. If your credit card is protected so that others do not know the number, date, and security code, then you could conclude that it is confidential. If the credit card has not been damaged so that the original numbers are still intact and if no one else claims it to be their card, then the card has integrity. However, if you cannot locate the card and do not remember the numbers, then you would be right to be concerned about the card's security. The aspect of security that is at issue in for the credit card is called *availability*. If information is unavailable, then it cannot be truly secure.

As another example of the role of confidentiality, integrity, and availability in security consider the use of cell phones and what it means for providers to offer a secure system. Certainly, we expect our phone conversations to be confidential so that only the people that we intend to hear our words are able to listen. We also expect the phone to provide integrity in the form of sound transmission that is accurate. We rely on the accuracy of the sound for data integrity. To some extent we also rely on sound for owner integrity when we identify the person on the other end of the call from the sound of their voice. Of course, owner integrity can be also be accomplished by looking at caller ID information for an incoming call. Supporting availability relies on the telephone provider's towers to be properly functioning, the transmission lines to have sufficient capacity, and phone coverage to be sufficient.

The three foundations of information security apply also to security in general. Criminal activity can be similarly classified. For example, impersonation and forgery are crimes of owner integrity; vandalism and murder are crimes of data integrity; theft is a crime of availability and/or confidentiality.

12.3 COMMON FORMS OF CYBERCRIME

Information security can be viewed as a contest between black hat attackers and everyone else. Unfortunately, the attackers have many advantages:

1. Black hats can play dirty; white hats must play by the rules.

2. Black hats choose which vulnerabilities to exploit; white hats must secure them all.

3. Black hats choose the time; white hats must be ever vigilant.

4. Black hats can invent new attacks; white hats only know about previous methods.

To make matters worse, it is surprisingly difficult for many computer scientists to think like an attacker. This is largely because a good software engineer is focused on making the software work. Exploiting software is an opposing view. In response to this white hat–black hat distinction, many software development companies hire outside companies or have separate teams to consider the security of their programs. These security specialists are responsible for what is known as *security assurance*, that is, ensuring that security systems are sufficiently strong.

The first step in security assurance, whether you are a security expert or an everyday computer user, is to understand black hats. In other words it is helpful for all of us to be able to think like an attacker. While there are a few attackers who are so skilled that they can create new methods of attack, the vast majority of attackers are far less sophisticated. This is because the Internet contains countless tools for mounting attacks that are freely available and easy to use. There seems to be no end to people, called *script kiddies*, with minimal computing knowledge but willing to use publicly available attacking software. Sometimes these script kiddies are malicious and sometimes they are merely curious experimenters, but they always pose a threat. Informed computer users can guard against script kiddie attacks. The first step to raise your guard is to understand the kinds of attacks that are typically used. In this section we shall survey the attacks, and Section 12.4 explores mitigations.

Sadly, the number of Internet security attacks has grown almost as rapidly as the Internet itself. A new form of crime, called *cybercrime*, describes these attacks, and the world's legal system is still trying to catch up with the attackers. The criminals include both technically gifted attackers and script kiddies. Most of their tools are computer programs, known as malware. Malware can be designed for many purposes, even including some that are white hat in nature.

Numerous exploits take place when an attacker manages to get your computer to execute the attacker's malware. Commonly, such exploits occur when the computer is "expecting" to receive some particular program or data, but the actual receipt has been corrupted by an attacker. Users are vulnerable whenever they download files, install software, or receive e-mail attachments, because any of these could contain *malware*.

Arguably the most widely known type of malware is known as a *computer virus*. Like biological viruses infect animals, a computer virus infects your machine. The damage that results from a virus infection varies from a simple warning message to erasure of an entire disk. Another characteristic common to both biological and computer viruses is that they are capable of spreading. Having infected one computer, a computer virus is designed to spread itself to other computers.

> One of the first known computer viruses to be distributed by the Internet is the *Morris worm*. In 1988 a Cornell University student, named Robert Morris, created the virus purportedly to measure the size of the Internet. Estimates suggest that 10% of the computers on the Internet at the time were infected, resulting in a cost of more than $10 million. Morris was the first person convicted under the federal Computer Fraud and Abuse Act.

As previously mentioned, malware typically results in a violation of data integrity. However, other common attacks can also result from breaches of owner integrity. Such attacks are referred to as *spoofing*. An e-mail message with a falsified return address or a network packet with an incorrect source address are examples of spoofing. Attackers also use spoofing when they mount *relay attacks* by making a transmission appear to originate from a different computer than their own.

Network sniffing is a common Internet attack for breaching confidentiality. A computer program, known as a *network sniffer*, is designed to capture network information and display it to the attacker. Network sniffing is possible in part because Internet messages can travel through wires and airborne transmissions is often transmitted in *plaintext* form, which means that they use simple data encodings like those discussed in Chapter 2. Plaintext data can be easily decoded by anyone able to intercept a message, including attackers.

Two less sophisticated forms of confidentiality breaches are known as *shoulder surfing* and *dumpster diving*. The names of these attacks are appropriately descriptive. Shoulder surfing occurs when someone observes confidential information. For example, an attacker might watch, or even capture a video, of someone typing his or her password. Dumpster diving is the act of stealing confidential materials that were intended to be discarded.

> Most corporate espionage goes unreported. However, in one well-publicized instance of dumpster diving from 2000 a private investigative firm hired by Oracle attempted to purchase garbage from Microsoft Corporation.

The Internet has also made attacks of availability commonplace. Many of these attacks are called *denial-of-service (DoS) attacks*, because the goal is to deny someone or some device the ability to send and/or receive information from the Internet. Sometimes these attacks are low tech, such as enlisting many of your friends to simultaneously send e-mail with large attachments to a selected victim in such a way that the person's mailbox is filled blocking further e-mail receipt. More sophisticated DoS attacks are mounted for the purpose of clogging Internet servers or routers to block information flow.

Even spam can be considered to be a weak form of DoS, although that is generally not the intent. The term *spam* refers to unsolicited e-mail and is considered to be the junk mail of the Internet. Any time you receive spam it wastes your time thereby at least slowing service, and large volumes of spam might even be considered to deny service.

Thus far, our discussion of confidentially has avoided a related concept: *privacy*. In security terms privacy is considered to be an asset. Each user has some right to privacy. You probably only share your cell phone number and your e-mail address with selected people and Internet sites. One small invasion of your privacy happens frequently when using Internet web servers in the form of *cookies*. A cookie is a file that is stored on your computer not by you, but rather by some web server you visit on the Internet. For example, if you are making an online purchase, the server used by the associated online retail company most likely uses a cookie to store your personal information. This cookie might store anything from nonconfidential things about your web browsing to a user name and password that you use to log in to the website. Cookies identify you to websites. The use of cookies allows a website to know who you are and perhaps tailor the information displayed to your personal interests. You can prohibit most web browsers from allowing cookies as a breach of your privacy, but doing so limits some of the functionality of websites.

> Just what should be private? This is an important ethical question. Clearly, telephone numbers and personal information could be useful tools for law enforcement in tracking terrorists. At the same time it is important to remember that Nazis used such information in the 1940s to persecute Jews.

If privacy is an asset, then the ultimate form of the asset is identity, and the associated crime is known as *identity theft*. An attacker commits identity theft by impersonating someone else. This form of crime does not require computers, but computers and the Internet have significantly contributed to its prevalence. Cybercriminals can search the Internet to locate things like Social Security numbers, bank accounts, and driver's licenses. Armed with such information they commit identity theft by opening new credit card accounts or checking accounts using the name and stolen information; this permits the attackers to spend money and leave the victim with the bill. Identity theft involves both a breach of confidentiality and owner integrity.

12.4 HOW TO SECURE? STEP 1: AUTHENTICATE

A core service of any security system is to *authenticate*. You are authenticating someone who knocks on your front door by asking them for their name or looking at them through a peephole in your front door. When a security system authenticates it verifies identity. Successful authentication mitigates identity theft and spoofing attacks.

In computer systems there are four categories (factors) that can be used to authenticate. Each factor represents a group of different techniques with a common characteristic. The four factors are based on

1. Something the subject knows

2. Something the subject possesses

3. Something the subject is

4. Somewhere the subject is located

FIGURE 12.2   Authentication tokens.

In these factors the word *subject* refers to the person, process, or device whose identity is being authenticated.

The most common mechanism for authenticating a computer user is a combination of a user name and a *password*. The password belongs to the first factor—*something the subject knows*—because users are responsible for knowing their own passwords and for keeping their passwords hidden from others. Therefore, security systems are often convinced of identity given a known user name and the proper password. A PIN used at ATMs is a second example of the something-the-subject-knows factor.

Physical security tends to rely more upon the factor called *something the subject possesses*. We possess keys that are used to open doors. We swipe credit cards to pay at the local grocery store. Increasingly, companies are using *authentication tokens* or *smartcards* as authentication devices (see Figure 12.2). These devices are either inserted into a reader attached to a computer or they are used to display codes that the user must type into the computer. In either case the user must have physical possession of the token or card in order to be authenticated.

The third factor, *something the subject is*, refers to a group of authentication techniques known as *biometrics*. Biometric authentication uses your actual human features. Forensic scientists have used fingerprints for many years to prove the identities of criminals. Some computer keyboards now incorporate small pads that scan a user's fingerprint and use it to authenticate. Other potential something-the-subject-is mechanisms that have been less frequently used include facial scans, iris scans, retinal scans, and voice prints. Biometric authentication is currently considered largely unreliable due to inaccuracies of current systems. For example, even fingerprints, the most accepted of the biometric mechanisms, was reliably spoofed in 2002 by a Japanese math researcher using just gummy bear candies. Despite current unreliability, considerable research is devoted to biometric authentication because of its convenience.

Security experts often limit the discussion of authentication to the first three factors, but in practice the fourth factor—*somewhere the subject is located*—is also used. For example, the credit card approval process sometimes depends upon the location where the credit card is used. Credit card companies maintain information about card-holder habits as well as where the card has been recently used. To guard against accidental authentication failure many people notify their credit card company before international travel. Similarly, you could make use of location as a part of authenticating e-mail you receive from an unknown sender by checking the country of origin, which is often included as a suffix in the e-mail address.

A subject's location is rarely sufficient to authenticate by itself, which is why the first three forms of authentication are considered to provide stronger authentication. To achieve even better authentication most security experts recommend the use of *two-factor authentication*. Two-factor authentication requires that mechanisms employing two different factors (usually excluding the fourth, location, factor as an option). For example, ATMs require both a credit card (something you possess) and a PIN (something you know). Authentication tokens also make use of two-factor authentication, because the user most possess the card and also know a personal code. Two-factor authentication is currently considered to be the gold standard for secure authentication.

Authenticating through the Internet is different than authenticating real-life humans. When you receive a web page you cannot use biometrics to authenticate the server that sent the page. Similarly, the distinction between something possessed and something known does not make sense in the context of e-mail. Therefore other means are used for authentication for electronic data on the Internet.

Computers are all assigned unique numbers when they are manufactured. These so-called *MAC addresses* identify each computer when it transmits data on a network. The Internet also uses another identifier known as an *IP number* to uniquely refer to each computer. IP numbers are assigned to computers by Internet administrators or by devices that distribute these numbers. Unfortunately, spoofing MAC addresses and IP numbers is relatively easy, so trusting these numbers from a received message is a weak method of authentication.

A reliable way to authenticate electronic data is to use what are known as *digital signatures*. A digital signature is a collection of bits appended to any electronic data. This collection of bits makes use of a clever encryption scheme in order to authenticate. Digital signatures are explained more fully in Section 12.7.1.

12.5 HOW TO SECURE? STEP 2: AUTHORIZATION

There are two entities that are key to security: *assets* and the *subjects*. By subjects we refer to the people, processes, or devices that access the assets. As previously explained, each subject is known by its identity and the purpose of authentication is to ensure the correctness of identity.

There are several identifiers (IDs) that are associated with identity. People are identified by name or some ID number, such as a social security number. Files are identified by a file name or a universal resource locator (URL). Computers on the Internet use MAC and/or IP numbers as their IDs, and Internet users are often identified by e-mail addresses (Figure 12.3).

An important characteristic of an identifier is uniqueness. Unfortunately names are often not unique. If you have a small group of friends or you operate a small company, then perhaps all your friends or employees have different names. In such situations, using names as an ID works well. However, if two different people have the same name, then the use of name as an ID leads to confusion. To ensure uniqueness, governments assign people numbers, such as the Social Security number or a driver's license number. Companies also commonly assign unique IDs to their employees.

Once your identity has been authenticated, it is time to worry about *authorization*. In any nontrivial security system, different subjects are permitted (authorized) different rights. For example, the president of the United States has certain privileges, whereas the party leader of the

IDENTITY
- Who are you?

AUTHENTICATION
- Can you prove
your identity?

AUTHORIZATION
- What privileges
do you have?

FIGURE 12.3   Identity, authentication, authorization.

People's Republic of China has a somewhat different collection of permissions. Even within the same organization the chief financial officer (CFO) is usually granted a different collection of access rights than the chief information officer (CIO).

When it comes to electronic information there are at least four different classes of authorization:

1. Read

2. Write

3. Own

4. Execute

A subject who is authorized to read a document is allowed to inspect the document, whereas a subject with write permission can alter or update the document. Ownership is somewhat unique in the sense that there is usually only one subject who owns an asset, and the owner is usually allowed to do such things as destroy, rename, or perhaps grant access permissions for the asset.

The execute class of authorization is unique to computer systems. Since many of the assets are actually executable software, we can separate the privilege of executing (or running) the program from the other three potential authorization classes.

Different subjects can be assigned different sets of authorizations. For example, you are allowed to read your checking account balance, but permission to change the balance is restricted to the bank. Users can change (write) their own passwords, and other users cannot. The files that you create are most likely owned by you, and you also have read and write permissions to those files. You can post (write authorization) messages to some but not all pages on a social media site, and you can view (read authorization) a different collection of pages from the site.

Often authorization is granted to *groups*, rather than individual subjects. For example, all members of a cell phone on a family plan may be able to view online the current status of the account. Similarly, all computer administrators have special privileges, such as the ability to create new users; creating a user is essentially a write authorization for the user database. Even the authorization to change authorizations is an authorization; often this is also a special privilege of an administrator.

Now that authorization has been discussed, we can offer a better definition for the word *breach*. A security breach is an event in which an attacker has gained unauthorized access to some asset.

12.6 ALL A MATTER OF RISK

It is probably better to use the word *strategy* than the word *solution* when it comes to providing security. Almost nothing can truly ensure total security. Do you completely trust the data integrity of your financial records? Do you really think that online retailers guarantee the confidentiality of your credit card information? Do you believe that government computers are immune from cyber attacks?

We speak about the security of our own home, but in fact homes are vandalized daily. Similarly, in the world of digital information it is impossible to be certain that confidential data has not been leaked or that the data we receive is not corrupted.

Risk is a combination of two things: the likelihood of a breach and the extent of resultant damage. It is best to think of security as a matter of probabilities. We adopt smart security practices in order to minimize risk or at least reduce the risk to acceptable levels.

For example, you have heard people say that you should use several symbols in your password. Given roughly 96 characters on the keyboard, if you choose a password that is one character long, then a computer algorithm only needs to check 96 potential passwords before being assured of discovering yours. If your password is two keystrokes in length, then the number of potential passwords grows to $96 \times 96 = 9,216$, but a few thousand possibilities is still child's play for a modern computer. If your password is 12 characters in length, then the number of possibilities skyrockets to over 10^{23}, which equates to lifetimes of computation by computers more powerful than we can imagine. The reason, as discussed earlier, is that the algorithm for discovering passwords with randomly selected characters has an extremely low probability of finding your password when there are too many alternatives.

Many times it is the choice of algorithms that determines the risk. If you select a password that can be found in a dictionary, or any other file, then you increase your risk. Cracking a password made of random characters requires an algorithm to check all possible combinations of characters. However, cracking a password that can be looked up from some dictionary only requires a computer program to check all of the words in the dictionary. Checking a few hundred thousand words can be accomplished in seconds by a computer program.

The informed user realizes that popular media is not always a reliable source of knowledge regarding technology and its risk. Despite what you might see on television or in movies, the probability of decrypting properly encrypted data is extremely low. (The next section provides more detail about encryption.) As with long random passwords, cracking soundly encrypted data can only be accomplished by algorithms that require too long to compute. However, your risk increases if you attempt to create your own encryption keys, use unreliable encryption software, or share private keys.

12.7 A FEW GOOD IDEAS

There are many strategies that have been designed by white hat computer scientists to mitigate information security problems. In Section 12.3 we examined mechanisms for authentication, which is a part of virtually all security systems. In this section we look at six other specific approaches commonly used to mitigate information security. Each subsection begins with the vulnerability, then explores an associated mitigation strategy to illustrate how computational thinking was applied.

12.7.1 Encryption

The first vulnerability is plaintext itself. The problem is that anyone can determine what characters or numbers are stored in a particular piece of computer data, because most data encoding formats are highly standardized and widely known common. Recall that the ways in which computer data encodes information was the main topic of Chapter 2.

Unfortunately, these encoding standards make it relatively easy for an attacker to discover the text of any data that has been intercepted. If an attacker can gain access to your files, e-mail, or other data, then the information contained by those data are easily discovered, violating any intended confidentiality. In addition attackers can create or corrupt data because they know how to formulate bit strings in the way that you interpret them.

A general mitigation scheme to overcome the encoding standard vulnerability is to obfuscate information by way of *encryption*. The idea of encryption is to scramble data (rearrange bit patterns) in a way that makes the data

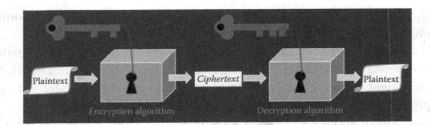

FIGURE 12.4   Two-way encryption.

essentially unintelligible to those who are unauthorized. The usual trick is to do this in a way that still allows access to those who are authorized.

Encryption requires the use of encrypting software, which is now a part of all computers. Encrypting a message can be described as transforming the plaintext version of the data into a scrambled (encrypted) form. The left portion of Figure 12.4 shows how two-way encrypting works. The original data in this figure is labeled plaintext and the name *ciphertext* refers to the encrypted form of the data.

Using *two-way encryption* is particularly effective for preserving confidentiality. Figure 12.3 shows that two-way encryption supports both a method for encrypting, as well as *decrypting*. Decryption is the opposite of encryption in the sense that a decryption algorithm translates ciphertext back into the plaintext from which it was encrypted. Using encryption to transmit e-mail (plaintext) confidentially requires that an encryption algorithm be applied to translate the e-mail before sending it in encrypted form (ciphertext). The recipient must then apply a decryption algorithm to translate your e-mail back to its original form (plaintext) to read it. But how is this possible? It turns out that the "magic" that makes this work is not just the algorithms for encrypting and decrypting.

Encryption and decryption algorithms are known to everyone, including attackers. The critical aspect that makes this possible is called a *key*. To encrypt a message, the encryption algorithm must be supplied an encryption key, and the encrypted message can only be decrypted given the corresponding decryption key. In other words, keys come in pairs; one key is for undoing the work of the other. (Sometimes, both keys are identical and other times they are different, depending upon the particular encryption algorithm.)

The name *key* is appropriate, because we use a similar idea for unlocking the door to our homes and automobiles. As long as you have the proper

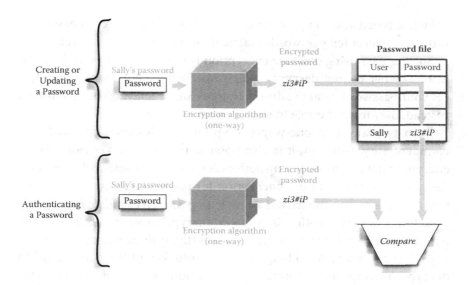

FIGURE 12.5   Using one-way encryption to store and authenticate passwords.

key to insert into a lock, then you are granted access. The main difference between house keys and encryption keys is that encryption keys are bit strings required for encryption and decryption rather than a physical key that unlocks some door.

To provide an extra measure of security for passwords, password files are usually encrypted in a somewhat different way, known as *one-way encryption*. One-way encryption uses special (one-way) algorithms that are designed so that while it is possible to encrypt, there is no known way to decrypt the ciphertext. We can conclude that no one should be able to decipher something that has been one-way encrypted.

One-way encrypting may seem useless. What good is it to store data that no one can decipher? However, one-way encryption leads to a secure mechanism for storing passwords. Figure 12.5 illustrates this. Figure 12.5 is a two-part drawing. The top portion diagrams the way that a user's password is encrypted for the first time. (The same procedure would be used when users change their passwords.) Notice that the password (in this case Sally's password) is encrypted and stored, along with the user name in a password file. The bottom part of Figure 12.5 depicts the password authentication algorithm. Every time Sally needs to be authenticated, she types her password again. The system cannot decrypt the password that has been stored in the password file, but it can encrypt the password that

Sally just typed and compare her encrypted entry to the stored password. If the two encrypted passwords match, then the password is correct.

Storing data using one-way encryption is so secure that even the password's owner cannot decrypt a password. This explains why if you forget your password, system administrators cannot tell you your password. Instead, they might be able to reset it to a new value.

Both two-way and one-way encryption are designed primarily for confidentiality. However, it is also possible that an attacker can corrupt encrypted data or spoof the owner/sender. Computer scientists have also devised a clever way to use encryption for preserving these two characteristics of security. The trick is an invention known as *public key encryption*. Public key encryption is two-way encryption where both keys of a pair are different from each other. Generally, it does not matter which key is used to encrypt, so long as the opposite key of the pair is used to decrypt. To implement public key encryption every subject must have their own pair of keys: one is called the *public key* and the other is the *private key*. Public keys are not hidden, but are given out freely, often to anyone who requests them. Private keys, however, are distributed to no one except the owner.

If Jane wishes to send a confidential message to Jose, then she first obtains Jose's public key, using it to encrypt the message. Since Jose is the only person with access to his private key, he alone can decrypt the message. This situation is depicted in the top part of Figure 12.6. Jane can use a somewhat different approach in order to preserve data integrity, shown in the bottom part of Figure 12.6. In this case Jane encrypts the message with her own private key, while Jose decrypts using Jane's public key. No one can encrypt a message that properly decrypts using Jane's public key, because she is the only one with access to her private key.

Unfortunately, the kind of data integrity that results from encrypting with your own private key is, by itself, not particularly useful. Consider what happens if Jose receives a message and decrypts it. How does Jose know that the message was sent by Jane and not some attacker? Jane's public key will decrypt any message; admittedly, an improperly encrypted message will most likely decrypt to meaningless text. However, unless Jose already knows the text of the original message was sent, he cannot be sure it is uncorrupted. In other words, before Jose can trust the message he needs first to authenticate who sent it.

Fortunately, if we take public key encryption one step further, it is possible to create a digital signature that ensures confidentiality, data integrity,

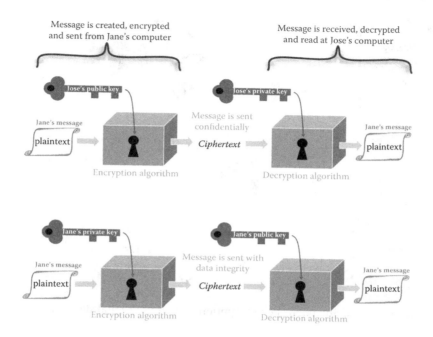

FIGURE 12.6   Two ways to use public key encryption.

and owner integrity (see Figure 12.7). The idea here is for Jane to append a second part, known as a *digital signature* and denoted "sig" in the figure. A digital signature is just an abbreviated digest of the plaintext that has been encrypted with Jane's private signature. The complete ciphertext consists of the encrypted plaintext together with the digital signature encrypted with the public key of the intended recipient (Jose).

Figure 12.7 also illustrates how security is maintained. The first step upon message receipt is to decrypt the ciphertext using Jose's private key. As described earlier, this ensures confidentiality, because only Jose is able to decrypt using his private key. This decryption returns the plaintext along with the digital signature. In order to guarantee integrity it is necessary to verify the correctness of the digital signature. Such correctness is verified by calculating a message digest from the plaintext using the same algorithm as Jane; in Figure 12.7 this calculated digest is called M2. In addition the digital signature (sig) part of the message is decrypted using Jane's public key, resulting in M1. Since only Jane knows her private key, when M1 is identical to M2 we can authenticate that it was Jane who sent the message. Furthermore, any attempt to corrupt the content of the message

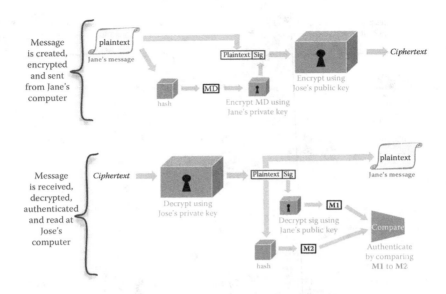

FIGURE 12.7   Using public key encryption including a digital signature.

would also cause M1 and M2 to differ. Therefore, whenever M1 equals M2 it must be that the message has data integrity and owner (Jane) integrity.

Digital signatures are widely used to secure Internet traffic. So long as private keys remain reliably private and public keys are available, a digital signature is a strong form of authentication. Digital signatures are used to ensure security when e-mail is sent in encrypted form and for most online purchase transactions.

Since encryption is based on sophisticated mathematical algorithms whose complexity is well beyond the understanding of most users, you might be wondering just what you need to do to use encryption, public keys, or digital signatures. Well, in the end you need to do very little, because most computer systems do the work for you. You will probably never create an encryption key, because these are generally created automatically for you. You probably have never seen your own digital signature, since they are also created as needed, and you are probably more secure without this knowledge.

The one level of frequent user involvement with encryption concerns *certificates*. Certificates provide a commonly used solution to the problem of how to obtain public keys. Our earlier discussion about public key encryption made the assumption that there is some way for public keys to be easily shared. However, making an encryption key publicly available is not as simple as it sounds. If you receive a public key by e-mail, you cannot be certain that the e-mail is not being spoofed. If you download a public key from a website, there is no guarantee that the website is authentic or that the key's integrity has not been damaged during transmission.

A certificate is nothing more than a file that resides on your computer that associates a public key with a particular identity. When your computer receives data or digital signatures that require decrypting using a public key, then the computer locates the proper certificate, extracts the public key, and automatically performs the necessary decryption on your behalf. When your computer system was first installed, the installation was initially preloaded with a few reliable certificates, many of which are for *certificate authorities* (*CAs*).

A CA is a company that serves as a source of certificates that can presumably be trusted. When your computer needs to communicate with some person or Internet device for which it has no certificate, your computer will generally check known CAs. A CA distributes certificates that are encrypted and signed (digitally) with the CAs own keys. Your computer must contain a certificate that authenticates the CA, and it can use this to authenticate any certificate stored by the CA.

There are many ways that you can arrange for your certificate to be validated by a particular CA. Typically, the CA has different ways to ensure that you are who you say you are. This might involve different levels of presenting various personal credentials and different costs are associated with different levels of security.

Sometimes your computer cannot locate a necessary public key. This happens when there is no stored certificate and when all of the trusted CAs fail to provide the necessary certificate. In these cases, some part of the encryption will not be completed. Often the user has to get involved in such cases, either by way of reading an error message or perhaps choosing what to do next.

12.7.2 Firewalls (Including Spam Filters)

Although the Internet provides countless sources of useful content, it also serves as an efficient attack vector. As soon as you connect your computer to the Internet, the chances for a security breach dramatically increase. Encryption provides one line of defense for data that enters or leaves your computer over a network, but a different approach is to use *firewalls*.

A firewall acts like a filter that allows some network traffic to pass through, while other traffic is blocked. The idea is to block traffic, especially those network messages intended to come into your computer, before they actually reach your computer. Hopefully, a firewall can block the messages sent by attackers. At the same time the firewall must allow desirable messages to pass through its filter.

Firewalls come in two forms:

1. Hardware devices

2. Software firewalls within computer systems

Some firewalls are physically separate units that are connected (usually with wires) and positioned between your computer and to the Internet. Your computer's only connection to the Internet is by way of the firewall device so that all Internet messages between your computer and the Internet must pass through, and therefore be filtered by, the firewall.

Most home computer users do not want the bother of a separate firewall device, so today's computer systems incorporate their own firewalls, sometimes called software firewalls. A software firewall inspects messages at the point of the network connection to your computer. A message from the Internet may still arrive at the network card inside your computer, but that message will immediately be filtered by the software firewall before it can actually do any damage.

One of the drawbacks of firewalls is that they only work when they are properly configured. If you administer your own computer, at the very least you must ensure that the firewall is turned on. You also have the option of specifying what gets filtered, but most users prefer not to get involved with the details of filtering.

Firewall devices tend to be more sophisticated than software firewalls, which is another reason most of us do not use them for home computers. That said, most companies do install firewall devices that protect the entire company network, including all of its computers. Such hardware firewalls are configured by network administrators, who understand the intricacies of network communications.

Firewalls are configured with *access control lists* (*ACLs*). An ACL is nothing more than a list of the types of messages allowed to pass through the firewall (permit authorization) and those messages that should be blocked (deny authorization). The difficulty with constructing these ACLs is how to specify which messages are which. One technique to do this uses the IP numbers of the computers that send and receive the messages. Another technique is to filter based upon the application inside your computer that is sending or receiving the messages.

Filtering based upon IP numbers is very common and is sometimes accomplished via *black lists*. A black list is a list of IP numbers that are known to have participated in attacks. The list needs to be distributed to your firewall by a trusted source, or stored within your computer and regularly updated by a trusted source.

Unwanted e-mail, that is, spam, can be blocked in a similar way using a *spam filter*. Spam filters are usually incorporated into e-mail client applications, the program that you use to receive and send e-mail. Sometimes the spam filter actually configures itself by recording which e-mail you quickly delete or by your marking certain messages as spam.

One problem common to both firewalls and spam filters is known as a *false positive*. A false positive occurs whenever a filter, or any other security system, erroneously detects an attack. In the case of filters, a false positive means that a message will be denied when it should have been permitted. In a sense, a false positive represents a security system that is overly protective.

12.7.3 Antivirus Software

You have no doubt heard the term *computer virus*. This expression is used to refer to a whole collection of different malware that can enter your computer. Viruses usually arrive through a network, but flash memory, CDs, DVDs, and external disks can also serve as virus vectors. The virus may consist of malicious software that your computer executes, or it may be other data that causes your computer's existing software to behave in unwanted ways. The final result may be nothing more than a humorous message displayed on your computer screen, or it may be as severe as erasure of your entire disk content.

The reason that the name virus is used is because, like viruses in the animal kingdom, computer viruses can spread from one computer to another. Of course, this spread requires some kind of vector. Just like sharing a glass of water can spread a human virus, sharing a CD can spread a computer virus.

Also, somewhat like human viruses, computer viruses do not necessarily infect every computer that is contacted. Often a computer virus only impacts a particular type of computer system or a particular application program. So computers using different systems or different programs will not be impacted by such specific attacks.

The most effective mitigation for viruses is probably *antivirus software*. An antivirus program executes on your computer and checks certain files or received data for potential viruses. These checks can occur automatically in the case of receipt of an e-mail attachment or manually when the user requests that files be scanned for viruses.

It may sound like antivirus software is similar to a firewall, and it is in the sense that both make use of lists to look for attacks. However, the lists used by antivirus software are more complex than those of a firewall. Antivirus lists store the particular bit sequence, often called a *bit signature* that is unique to a virus. This is almost like a genetic encoding for a digital virus. When the antivirus software detects a bit signature that matches a known virus, then the file containing the bit pattern is considered to be infected with a virus. Such infected files may be automatically deleted, or sometimes the user is prompted to determine another response.

The name *antivirus* unfairly describes this mitigation technique, because this software can detect more than viruses. Technically, a virus is capable of spreading itself. However, there are other forms of malicious software that arrive at your computer in similar ways to viruses. All of these kinds of malicious software can be mitigated using antivirus programs.

Antivirus software can produce false positives, but a more common problem is that only known malicious bit signatures can be detected. Viruses can be spread in milliseconds, but discovering unique bit signatures for a newly released virus can require hours or days, and distributing this to your antivirus software can take even longer. So while, antivirus software can effectively mitigate known viruses, new viruses often spread undetected.

12.7.4 Software Update

By now you should be getting the picture that techniques like firewalls and antivirus software are only as good as the lists and tables used to differentiate good from evil. Failing to keep ACLs, black lists, and known malicious virus signatures up to date is a vulnerability.

The need for regular computer updates is also fueled by another vulnerability: *software faults*. The simple truth is that virtually all computer programs of sufficient size contain faults. Some of these faults cause the software to fail to perform properly on occasion. Many such faults have security implications. In August 2003 the brown out that impacted the northeastern United States and Canada for several days was the result of a software fault that took weeks for the electric company to completely correct. Imagine the potential damage that could result from software faults in airplane guidance systems or sensitive medical equipment.

Despite the many significant accomplishments of computer software, it is still the case that software faults have proven to be an inevitability. If we cannot eliminate them, we must plan for them, and one way to do this is *software update*.

Responsible firms that create commercial computer software monitor their programs to detect problems. When a problem is severe (and security-related faults certainly qualify as severe), the company is obligated to correct the software fault and release a new version of the program without the detected problem. These new software programs are delivered to your computer over the Internet by way of software updates, also known as *software patches*. Some computer systems automatically check for such updates and others require that you manually request an update check. Updates need to be installed in a timely fashion, and they should be performed immediately when marked "critical."

12.7.5 Backups

Despite all your good intentions and careful adherence to strong security, the unthinkable happens: your machine is attacked. Perhaps malicious software has infected your computer so that the integrity of your files is not trustworthy. Perhaps using the machine in its current form exposes everything that you thought was confidential. In any event, your computer has been breached.

There is one mitigation technique that can assist even for this case of a serious breach. When all else fails, computer users resort to a *detect and recover* approach. First, the user must detect that the breach has occurred and then in response must recover key data.

A recovery strategy relies on users performing sufficiently frequent *backups*. A backup occurs when you copy files to some media outside your computer, usually an externally connected hard disk or a server. Modern computer systems include programs to assist you with backup, but the creation of backups almost always needs to be activated by the user.

Backups are also used to compensate for hardware failure. For example, if your computer's processor or hard disk quit working, you should still be able to recover the machine's files so long as you have made a recent backup. This same backup procedure is similarly effective at recovering from a security attack.

12.7.6 Log Files

Still another risk involving computer systems has more to do with our legal systems than our computers. It is always possible that computer data can play a role in matters of law. Perhaps you are involved in a lawsuit concerning ownership of particular data stored on some computer. Perhaps a computer has been used to perpetrate some crime. Perhaps we are trying to trace a terrorist through a collection of computers. All these cases involve vulnerabilities that might depend upon a kind of forensic data gathering technique known as *log files*.

Just as the name implies, log files log (or record) various events during the use of your computer. Events that can be logged include such things as a user logging in, a file being created or read, and an e-mail message being sent or received. Computer systems can, and often do, record many events. A certain amount of logging occurs by default, but users can also change system settings to fine-tune what is included in log files. The default settings are probably sufficient for most personal computer users, but you should be aware that this information is available.

12.8 GOOD STRATEGIES

In the previous section we examined six classes of vulnerabilities and six general mitigation strategies. In this section we survey selected security principles and best practices. All of these represent the collective wisdom of many computer scientists over the period of several years, the computational thinking of many.

12.8.1 Secure the Weakest Link

In an earlier discussion we examined the fact that security is always a matter of risk. The probability of certain threats is greater than others, so it just makes sense to mitigate those with greatest likelihood for a breach. We call this securing the weakest link(s).

One of the reasons that firewalls and antivirus software is effective is that script kiddies vastly outnumber professional attackers. A script kiddie almost always uses attacks and attack vectors that are well known; known attacks are those best mitigated by firewalls and antivirus software.

Whether we like it or not, the weakest link in many security systems is us, the users. Humans are vulnerable to a class of attacks known as *social engineering* that are designed to cause us to behave in ways that break down security. For example, an attacker may impersonate a high-ranking member of your company and ask for your user password. Or perhaps the attacker claims to be a member of the IT (information technology) staff and asks you to log in to your account so that maintenance can be performed. These kinds of social engineering attacks have proven successful mostly because people want to please their superiors or want to be helpful.

Organizations use two primary tools for mitigating against social engineering attacks:

1. *Security policies*

2. *Security education*

Smart organizations have established policies and procedures involving security. These policies might refer to things like logging off when you are away from the computer, and procedures for classifying and handling confidential material. Policies might also describe who has authorization for what and procedures for working with people you do not know. Policies like these are only useful if they are known and understood. Therefore, today's organizations typically include information security training as a regular part of their education program.

12.8.2 Reduce the Attack Surface

Securing the weakest link may be a good idea, but what about the other threats? Computer professionals refer to the extent of your vulnerability, including potential attack vectors, as the *attack surface*, and it is always a good idea to reduce this.

For example, when you are typing a password or viewing a confidential document on your computer, you need to be aware of who or what has a view of your computer. These days most cell phones can video your typing or photograph your display. In such cases you might be able to reduce your attack surface by simply closing an office door or waiting until someone walks to a different location.

Another simple way you can reduce your attack surface is by disconnecting your computer from the network when you are not using the Internet. Similarly, you could turn off your computer when it is not being used.

Unnecessary data should also be disposed of securely. It may seem like an obviously bad idea, but many people still keep passwords written somewhere. Furthermore, you should be aware of all files that are confidential and delete such files when they are no longer needed.

Even disposing of an old computer can increase your attack surface. Data stored on magnetic devices, such as hard disks, is not easily erased. Just because you have deleted a file does not mean that the data in that file is truly unrecoverable. Sometimes delete commands merely mark a file as deleted and do not actually remove any data from the disk, and even data that actually is deleted can leave residual magnetic information. In fact highly secure installations warn that the entire disk surface should be overwritten with 0s then 1s then 0s then 1s several times before data can be considered to be safely gone.

Finally, the earlier mentioned mitigation strategies, such as encryption, firewalls, and antivirus software, all contribute to an extent in reducing the attack surface. Again, it all depends upon how much risk you are willing to accept. For example, encrypting files can be time consuming so most of us do not unless the data in those files is highly confidential.

12.8.3 Defend Deeply

Hopefully, you are now convinced that no mitigation technique is foolproof. Even encryption is susceptible to the use of an insecure algorithm or insecure use of a strong algorithm or just plain a lucky guess. For this reason users are strongly encouraged to practice security in layers.

Even in medieval times the concept of multiple layers of defense (defending deeply) was well understood. A king's castle was often surrounded by a moat; but if the attackers managed to get past this water, they had to confront the castle walls. Once inside a wall the king's guards took charge of the next line of defense.

We could all do well to think of our computers as a kind of personal castle. If the first line of defense is a firewall, then antivirus software may be our second line. The standard computer system user authentication and authorization provides yet another layer, as does encryption of data transfer and file encryption. Guarding against social engineering attacks add still more depth to our security systems.

If none of our defense layers is impermeable, at least we can be comforted by the fact that circumventing all of our different countermeasures will be difficult.

12.8.4 Compartmentalize

Defending in depth seems to imply that we treat all assets as a group. However, it is often smarter to subdivide assets into separate compartments and secure the compartments independently. The hulls of large ships are certainly compartmentalized, so that flooding a single compartment does not necessarily mean others are also breached.

Data security can similarly benefit from compartmentalization. Several companies have reported that their online servers had been successfully attacked and that these servers stored confidential credit card information from their customers. It might have been wise for these companies to use multiple files, systems, or servers to partition their customer credit card information so that compromising one group of customers did not necessarily compromise all.

The concepts of data authorization are at the heart of compartmentalization. We can define access groups that form a collection of individuals who all share the same access rights, a kind of compartment. Confidentiality policies turn these access rights into a set of practices.

Arguably the best example is the *need-to-know policy*, used by many organizations concerned about confidentiality of information. As its name suggests, individuals should only be told as much as they need to know. Suppose your company has discovered a new drug that will eliminate malaria, but you do not want your competitors to know. If the company adheres to a strong need-to-know policy, then only the people who simply must be aware of the new drug know anything about it. This creates a compartment while also following an earlier principle by minimizing the attack surface.

More complicated confidentiality policies are used by governments and militaries. These policies define asset classification levels (such as unclassified, classified, and top secret) and subject clearance levels used to determine which subject has access to which assets.

Preserving availability can lead to another kind of compartmentalization. Rather than dividing the data into smaller groups, it is sometimes best to think of duplicating the entirety of the data. This idea is called *mirroring*.

Organizations with truly critical data often utilize mirrored storage that duplicates all stored data in more than one computer. For example, a medical clinic might store all patient records in files on a server in the home site of Chicago, while at the same time storing identical files in a mirroring server in Los Angeles. This mirror site might increase the attack surface but is often an acceptable risk for the potential gain in availability should either site fail.

12.8.5 Trust Reluctantly

All security ultimately depends upon trust. We trust that our user password has not been compromised. We trust that our software maintains the integrity of our data. We trust that a certificate authority is reliable.

Still, it is wise to be somewhat skeptical about who or what to trust. Most of us know enough not to trust e-mail that was sent from an unknown individual. Furthermore, you should not download software from just any web server. Before making a purchase on the Internet, it is a good idea to consult one of the websites devoted to rating online retailers.

Social engineering attacks only succeed due to misplaced trust. You should always be suspicious when a stranger tries too hard to befriend you or tries to pressure you into quick action. Even a uniform can be the sign of an attempt to improperly gain your trust.

One attack form that is all too common today is referred to as *phishing*. The concept of a phishing attack is to acquire confidential information by spoofing a website. Often the attack begins with a link sent through e-mail or a text message. Following the link might take you to a fake website that looks like some commonly used website. The goal is often to convince you to log into an account on this phishing website, at which time your user name and password have been captured and the attacker can use them to log into the real website.

One countermeasure that mitigates against trust-based attacks is a *callback*. The name *callback* comes from telephones. When you receive a phone call from someone claiming to be from your bank, or other important source, you can use a callback to improve security. To callback, you do not say anything important to the person who made the call, but rather you call (back) to the bank using a trusted phone number. A callback works because it puts you, not the potential attacker, in control of initiating contact.

Callbacks are equally useful for Internet phishing. Instead of following a link embedded in an e-mail message, you could type the URL for the intended website into your web browser, thus calling back to a trusted address. For e-mail, it is more secure to avoid the reply button and callback by creating a new e-mail message and typing the trusted e-mail address.

12.8.6 Use Open Software

A relatively recent but relevant development for computing is a movement known as *open software*. The word, *open* means publicly available in this context. There are three forms that this openness can take: (1) free software, (2) open source, and (3) open design. Free software is neither inherently more or less secure, but the other two forms of open software can result in security benefits.

When a program is called "open source" that means that the original code is publicly available. Generally, this means that the software is also free. The Free Software Foundation (www.fsf.org) actively promotes the concept of open source.

From a security point of view the advantage of open source is that the software has most likely been examined by far more people. Since most computer scientists qualify as white hats, not black hats, having more people look at the code is beneficial. The GNU operating system and other variants of Unix, including Linux, are open source, which is one of the reasons claimed for why these systems are widely considered to be more secure than Windows, which is proprietary.

Sometimes it is not necessary to provide open code but just to be transparent about the design of software. *Cryptographers*, those computer scientists and mathematicians who invent encryption algorithms, have commonly practiced a form of open design by sharing their algorithms. Any cryptographer will tell you that an encryption algorithm must stand the test of time before it can be considered to be useful. What this means is that once the new algorithm is released, then other cryptographers attempt to discredit it by cracking encryption keys. Only after the cryptographic community has had sufficient time to attempt to unsuccessfully break the algorithm is it considered to be a secure form of encryption.

12.9 WHEN WILL YOU EVER USE THIS STUFF?

If you think technology can solve your security problems, then you don't understand the problems and you don't understand the technology.

—BRUCE SCHNEIER

We are exposed to countless forms of security attacks these days: phishing attacks, spoofing, spamming, even identity theft. Our only hope is that white hats create mitigation strategies at a similar rate to the black hats' newest attacks. But all of these mitigations are useless unless we know and understand the basics of both the attack and the defenses.

REFERENCE

1. "2010 Cybersecurity Watch Survey: Cybercrime Increasing Faster than Some Company Defenses," www.cert.org/archive/pdf/ecrimesummary10. pdf. Conducted and compiled annually by *CSO Magazine* in cooperation with the United States Secret Service and US CERT, 2010.

TERMINOLOGY

access control list (ACL)	cookie
antivirus software	cybercrime
asset	data integrity
attack surface	decrypt
attacker	denial of service (DoS) attack
authentication	digital signature
authentication token	dumpster diving
authorization	encrypt
availability	encryption
backup	execute authorization
biometric authentication	exploit
black hats	firewall
black list	false positive
breach	group authorization
callback	identity
certificate (for authentication)	identity theft
certificate authority (CA)	information security
ciphertext	IP number
confidentiality	integrity

log file

MAC address

mirrored data

mitigation

need-to-know policy

network sniffer

one-way encryption

open software

open source

own authorization

owner integrity

password

phishing

physical security

plaintext

privacy

private key

public key

public key encryption

read authorization

relay attack

risk

script kiddies

security assurance

security system

shoulder surfing

social engineering

software fault

software patch

software update

smartcard

spam filter

spoofing

threat

two-factor authentication

two-way encryption

vector (security)

virus

vulnerability

white hats

write authorization

EXERCISES

1. Following is a list of specific mitigation strategies discussed in this chapter. For the threats explained in parts (a) through (d) list all of these mitigation strategies that can significantly improve security.

   antivirus software

   callback

digital signature

file backup

firewall

log files

one-way encryption

two-way encryption

spam filter

a. You are a reporter working in a country known to harbor terrorists and you need to exchange e-mail that preserves both confidentiality and integrity with your home office.

b. You work for the government and your machine stores information that is highly classified. You do not use this computer for e-mail, but it is connected to the Internet. Not only are you expected to maintain confidentiality of the classified information, but it is also expected anyone attempting an attack must be caught and prosecuted.

c. You are writing the computer program to store personal identification numbers (PINs) for an ATM company.

d. You work for a financial company and your boss frequently e-mails you instructions to make large, expensive stock purchases and sales. If you get the instructions wrong, you could lose your job.

2. Suppose you are responsible for securing many paintings from famous masters. These paintings are maintained in storage when not on loan to some museum. Explain a specific way in which you might take advantage of each of the following security principles to make these paintings more secure.

   a. Compartmentalize

   b. Defend deeply

   c. Reduce the attack surface

   d. Secure the weakest link

   e. Trust reluctantly

3. List all of the following that would be considered two-factor authentication.

   a. Using a fingerprint and a password to log on to your computer

   b. Requiring a key and an ID card (read by a swipe card reader) to open an office door

   c. Requiring that you know where to find the secret lever and also a magic word to open the sorcerer's lair

   d. Performing both a facial scan and an iris scan before allowing passengers to pass through airport security

4. Label each of parts (a) through (e) with all of the following authorization types that are required in order to accomplish the specified task: *read, own, execute, write.*

   a. You wish to create and empty file.

   b. You have a directory of employees, and without altering the rest of the directory you merely want to add a new employee to the end.

   c. You need to understand a file of instructions but also to edit any misspellings in the file.

   d. A file, consisting of your e-mail client program, needs to be run.

   e. You wish to delete a file.

5. You have designed a new security plan for your company. Rather than hide the plan, you decide to share it with all company employees. Of course, they may expose company assets to insider attacks, but you are able to justify this using one of the security principles from Section 12.8. Which principle would you use and what is your justification?

6. Make a list of specific suggestions that would be good advice for anyone using a personal computer who is worried about identity theft.

7. How many keys in total are needed for four people to be able to communicate securely, that is, each person can communicate securely with any other person.

8. When you answer your cell phone, you listen intently to the voice on the other end of the call. Are you trying to authenticate or authorize?

Index